Change in the Weather

Change in the Weather

Philip Eden

Continuum
The Tower Building
11 York Road
London SE1 7NX

80 Maiden Lane
Suite 704
New York
NY 10038

www.continuumbooks.com

First published 2005
Reprinted 2006

British Library Cataloguing-in-Publication Data
A catalogue record for this book is available from the British Library.

ISBN 0-8264-7973-1 - Hardback
ISBN 0-8264-8029-2 - Paperback

Typeset by Kenneth Burnley, Wirral, Cheshire
Printed and bound by Cromwell Press Ltd, Trowbridge, Wiltshire

Contents

Foreword

It is extraordinary what a sadistic old crone Mother Nature is when you think about it. Even in a comparatively benign climate like our own, we have gallons of water thrown at us at regular intervals, we can be blown halfway across the country by winds gusting to 100 mph or more, blankets of fog may hide practically everything familiar from view, and we run the risk of being blasted into the middle of next week by a million-volt bolt of lightning. And on occasion she will even launch an artillery barrage of solid ice missiles at 50 or 60 mph – in other words, hailstones – or utterly transform the landscape under a six-inch layer of snow.

Bad weather has always been part of the British scene, yet in recent decades something odd has been happening to the way we perceive our climate. These days, when we get a bout of unusual or extreme weather, especially if it impinges adversely on our day-to-day lives, we ask what has gone wrong, we point the finger at our service providers – from weather forecasters to those responsible for gritting the roads, running the railways, or managing our rivers – and we demand a scapegoat. Every time the weather turns nasty we seem to think it is something brand new. But it's not.

Some of the blame should lie with ourselves. As a society, most of us now live in centrally heated homes, work in environmentally controlled offices or factories, and drive around in comfortable motor cars which protect us from the wind and rain. We are becoming divorced from our environment. We don't notice the weather unless it actually penetrates our cocoon and disrupts our daily routine. When that happens we get annoyed, rather than cope with the problem and simply get on with our lives as our parents and grandparents did.

Television, radio and the newspapers also bear a large measure of responsibility for changing our perceptions. Journalists acknowledge a phenomenon in their trade called 'adjectival inflation' (although the problem is not restricted to that part of speech): every accident is a serious accident, every death is a sad death, every breakthrough is a dramatic breakthrough, every revelation is a sensational revelation, and so on. This works splendidly with weather events too: a wind strong enough to lift roof tiles and break branches off trees becomes an '80 mph hurricane', a routine thunderstorm becomes a 'freak electrical storm', a few flakes of snow turn into a 'Siberian blizzard' or simply 'Arctic hell', and 12 inches of floodwater collecting under a railway bridge in Neasden becomes 'London's dramatic flash flood'. We are therefore bullied into believing that a weather event which may happen half a dozen times in any given year is a freak of nature which demands some sort of explanation.

In an era when we are rightly worried about the possibility of serious changes in our climate, it is essential that we reconnect with the climate we have at the moment. We cannot accurately assess what a different sort of climate might do to us in the future if we have forgotten what our very own climate has done to us in the past. We only have to look at the Indian Ocean tsunami of Boxing Day 2004 to see what happens when a credible threat is perceived to be too small or too distant to worry about: this catastrophe that snuffed out more than a quarter of a million lives happened a month after a conference of interested countries decided that the risk was so small that establishing a tsunami warning centre in the region would be a waste of money.

The main purpose of my regular columns in the *Daily* and *Sunday Telegraph* that have been collected together in this volume is to remind us all of the range of weather events, some common, some rare, that we have experienced in the British Isles in recent generations, and to show that, however bizarre something may seem at the time, there is always a scientific explanation for it. Only in this way can we have proper historical and scientific contexts in which to place the sequence of interesting and unusual meteorological phenomena that we shall undoubtedly experience in the coming years.

Philip Eden
March 2005

Chapter 1

The rain-making machine

Introduction

Seen from space, the Earth looks serene, its blues, greens, whites and browns combining in gently changing patterns. From the ground, the atmosphere can also be utterly peaceful, as on a sunny spring morning, but it can also generate extreme violence as in a typhoon, a tornado or a thunderstorm. The weather is the state of the atmosphere in a given place at a given time, and the vagaries of the weather influence every aspect of our lives. The weather has a say in the sorts of houses we live in, the character of the towns and villages we inhabit, the kinds of animals and plants raised on our farms, the sort of food we eat, and the agricultural methods we have developed over the centuries to produce that food. Our lives and lifestyles are so intricately interwoven with the behaviour of the weather that it is hardly surprising that we try so hard to predict what it is going to do next.

The atmosphere is really a giant heat engine, and the weather is its output. The interaction between heat energy and water vapour is, in fact, responsible for practically all of the day-to-day weather that we experience. The source of heat energy – the sun – is external to the Earth and its atmosphere, but the moisture is supplied 'in house', and no significant water vapour is gained from or lost to outer space.

Water in its various forms travels between the ground and the air in a never-ending two-way traffic, and this is known in the business as the *hydrological cycle*. The downward traffic takes the form of precipitation. This is not just rain, but snow and sleet and hail as well. Some of this falls directly over the oceans, but the greater part of the precipitation that falls over the land also finds its way back to the oceans via rivers and streams. The upward traffic is the result of evaporation and transpiration: evaporation from the oceans, seas and lakes, and from the rain-soaked land too, and transpiration from the plants that make up forests and grasslands, meadows and crops, gardens and parks.

Another fundamental element of the weather machine is motion. Even if the Earth did not rotate on its axis, air currents would still exist in the atmosphere, simply because warm air is less dense than cool air. Bubbles of warm air rise and are replaced by surrounding cooler air. This is basically what wind is – air in motion. However, the rotation of the Earth injects large additional quantities of energy, resulting in the mass movement of air across oceans and continents and between the tropical and polar regions. Those quasi-circular storms, from the vast depressions of the Atlantic and Pacific Oceans to the small but violent tornadoes, all owe their existence to the power supplied by the Earth's rotation.

What are clouds made of?

Our spacious skies provide a never-ending focus of interest, a daily pageant of sunshine and shade, of blue sky and clouds, for those of us who care to cast our eyes upward. Sometimes the sky is static and featureless, at other times changing from moment to moment, the interplay of sunlight and shadow on the Earth beneath stimulating the senses. Even urban scenery, drab and depressing under grey skies, can take on a new lease of life when the clouds break and sunshine streams through, glinting on windows and puddles, and all of a sudden everything looks freshly painted.

Clouds form when invisible water vapour in the atmosphere condenses into visible water droplets. All air contains some water vapour, but a given volume of warm air can support more water vapour than the same volume of cold air. When a flow of air is forced to rise – over hills or mountains, or over a large mass of cold stagnant air, or when it becomes buoyant following contact with the warm ground or ocean beneath – it moves into a part of the atmosphere where the pressure is lower. As its pressure falls the air will expand, and as it expands it will become progressively cooler, until it reaches a point where it is no longer able to hold on to all of its moisture. The excess water vapour condenses into droplets or into ice crystals, and our cloud is born.

A country person with years of experience of watching the weather would probably recognize two kinds of cloud – rain clouds and fair-weather clouds. Essentially, rain clouds are ones with strong vertical updraughts of air which hold water droplets and ice crystals aloft long enough for them to grow quite large. Eventually, though, gravity will win, and the droplets or crystals become heavy enough to fall through the bottom of the cloud as precipitation – rain, snow, hail, and so on. Much depends on the temperature and humidity of the air outside the cloud, but as a general rule the shallower the cloud, the less chance there is for raindrops – or snowflakes or hailstones – to form and fall to the ground.

A country person also uses clouds to forecast the weather over the next few hours, and, if skilled at the task, may be able to do so with a fair degree of accuracy. In past times, before Luke Howard introduced a scientific classification of clouds, familiar cloud types were likened to other parts of the natural environment: mares' tails, mackerel sky, heaps of wool, flocks of sheep, piles of fleeces, goats' hair, rocks and towers, water wagons, high fog. Some of these are so accurately descriptive that they remain in use today. The sixteenth-century mathematician and astronomer Leonard Digges is perhaps best known for his pioneering work on the design and construction of telescopes, but he also turned his hand to meteorology, and his *Prognostications to Judge the Weather*, published in 1555, illustrates how important clouds were to weather forecasting in those days:

> If thick clowdes resembling flockes, or rather great heapes of woll, be gatherid in many places, they shewe rayne. Also when grosse, thicke, darke clowdes, right ouer the northe part, or somewhat declynyng to the west, ar close with the earth, immediatly folowyth rayne. If they appeare lyke hylles, somedeale from the earth, a good token of weather ouerpassed. Blacke clowdes, signifie rayne: white clowdes apperyng in wynter, at the Horizon, two or thre dayes together, prognosticate colde, and snowe.

Shakespeare occasionally cast his eye to the sky, too, as the following passage from *Antony and Cleopatra* vividly demonstrates:

> Sometime we see a cloud that's dragonish.
> A vapour sometime like a bear or lion,
> A tower'd citadel, a pendant rock,
> A forked mountain, or blue promontory
> With trees upon't, that nod unto the world,
> And mock our eyes with air: thou hast seen these signs;
> They are black vesper's pageants . . .
> That which is now a horse, even with a thought
> The rack dislimns and makes it indistinct
> As water is in water.

Meteorologists see clouds as part of the global weather machine, each a small but important cog or wheel, demonstrating the mass movement of air from ocean to continent, say, or the transfer of huge amounts of energy from one latitude to another. They all play a part in the delivery of different kinds of weather, and an experienced modern-day weather forecaster, if dropped into a field with no access to data streams and computer models, would be able to make an accurate forecast for three or four hours ahead – and have a pretty good stab at the next twelve hours – simply by studying the sky.

Rain, rain, go away . . .

Everyone hates to get caught in a downpour. But we need rain to stay alive, to grow food, to fill our reservoirs and to keep the countryside green. We have to pay a price for this essential resource: at a trivial level we might get wet when it rains, or sporting fixtures may be abandoned; more seriously we may find our streets impassable as floodwaters rise, and occasionally we may have to cope with the extreme discomfort of having our homes flooded out as well.

Of the various meteorological elements, rain is easily the most important. Our supply of water for drinking and bathing depends on rain. So does water for industrial and leisure activities, and for growing crops and raising livestock. In areas where rainfall is not sufficient for arable farming or for rearing animals, water may be imported by irrigation.

A shortage of rain brings a clutch of problems. Springs and wells fail; reservoirs dry out. The green countryside turns brown, and the supply of water to every home – taken for granted most of the time – may be rationed. The yield of cereal and root crops will decline, and in extreme cases the harvest may fail altogether. In marginal societies, that could well lead to famine. An excess of rain brings different problems, washing away crops, damaging property, cutting transport arteries, and disrupting the everyday lives of country folk and town dwellers alike. Flooded rivers may overtop their banks, drowning livestock, and making farmland unusable for months. In wealthy countries such events are rarely catastrophic, but in regions where farmers consume what they produce, stocks of food for the population as a whole may be minimal, and the consequence will again be famine.

Rain always falls from clouds, but not all clouds produce rain. Some banks of cloud may last for days on end without releasing the tiniest speck of drizzle, while others may produce a torrential downpour within a couple of hours of appearing. Particular atmospheric conditions are required for ordinary-looking clouds to grow into rain-making clouds, triggering the processes whereby tiny cloud particles grow into drops big enough and heavy enough to fall to the ground.

Broadly speaking, ascending air within a cloud is what distinguishes one that produces rain from one that does not. The rain cloud also needs billions of 'condensation nuclei' – tiny particles of dust, salt, soot and pollen upon which the water droplets grow. If our atmosphere were utterly pure with no condensation nuclei, life on Earth (if it existed at all) would be very different because there would be no clouds, no rain, no major rivers, and all precipitation would take the form of dew and frost.

There are two principal ways in which these cloud particles grow into drizzle droplets, and ultimately into raindrops that are heavy enough to fall to the ground. One, the 'coalescence process', involves minute droplets colliding with each other to form larger and larger drops, eventually becoming full-sized raindrops. Since this can happen in clouds with no ice, it generates much of the rain that falls in the tropics. In any particular cloud, there is a range of droplet sizes, probably due to the different sizes and characteristics of the condensation nuclei. Where the rising currents of air are especially strong – as they are in cumulus and cumulonimbus clouds – the droplets are carried further and further upwards.

The large ones, however, are carried upwards more slowly than the smaller ones. This will result in repeated collisions between water drops. This process is more efficient the deeper the cloud becomes, and the stronger the rising air currents become. When the bigger droplets finally reach the top of the cloud, they will no longer be supported by the updraught and will therefore fall back through the cloud, growing all the time as they absorb countless other water particles.

The other way in which rain is created is known as the 'Bergeron–Findeisen theory', after the Swedish and German scientists who first described it, or more simply the 'ice-crystal process'. Because it requires the cloud to contain both supercooled water droplets and ice crystals, it is the main process found in rain clouds in temperate and polar regions. At any given temperature, water droplets evaporate more quickly than do ice crystals. When supercooled water droplets – that is, water still in liquid form even though the temperature is below 0°C – and ice crystals are found together in the same cloud, moisture evaporating from the water drops condenses on the ice particles which therefore grow rapidly. The net effect of this process is that the ice crystals grow at the expense of the water droplets. The larger ice crystals coalesce with smaller ones, eventually forming snowflakes. These continue to grow until they become too heavy to be held up by the rising air currents, and they fall through the cloud. If the air is cold it will reach the ground as snow, but in temperate latitudes snowflakes may start melting in the lower part of the cloud or between the cloud and the ground, resulting in rain.

The stronger the rising air currents in the cloud, the larger the snowflakes or water drops become, and that is why the biggest raindrops are normally associated with giant thunder clouds. By contrast, when the upward motion of the air in a cloud is very sluggish, and the cloud is relatively shallow, any precipitation that does make it to ground is likely to be in the form of drizzle.

The rain-shadow effect

'Orographic enhancement' and 'rain shadow' are phrases often used by meteorologists in their day-to-day work. Some may regard both as solecisms (orographic is from the Greek *oros* meaning 'mountain' and *graphein*, 'to write'), but taken together they well describe the phenomenon whereby much more rain falls on the windward side of a mountain range than on the leeward side. Most years provide several examples of orographically enhanced rainfall, but the first week of February 2004 produced the best example of these two complementary effects that we have seen in the UK in recent years. It was all thanks to a persistent south-westerly airflow delivering warm and moist air that had originated in tropical latitudes of the Atlantic Ocean, somewhere to the south or south-west of the Azores.

Both effects are best observed when the wind blows consistently from one direction, delivering very moist air following a long journey across the ocean. The air is forced to rise over a range of hills in its path, and as it rises it becomes less dense because the barometric pressure aloft is lower than it is at sea level – that is, there is less weight of air above. The laws of physics tell us that when the density of an air-mass decreases, its temperature will also decrease, and the cooler it becomes the less moisture it can support. If the air is already saturated when it reaches the mountains, the excess moisture will condense into cloud-droplets, quickly producing rain, and if the airflow persists for several days large quantities of rain are likely to fall over these windward slopes.

Things are very different on the leeward side of the mountains. Thanks to the prolonged rain on the windward slopes, the air-mass has now lost much of its moisture, and as the winds descend the lee slope the air becomes denser and therefore warmer. As it warms up, its capacity to hold moisture increases again, so it is no longer saturated. The mechanism which produced the persistent rain on the windward slope is now switched off, the rain stops, and the clouds dissipate. Indeed, the air may dry out so efficiently that its relative humidity can drop to 30 per cent or below.

Some figures for that week in February 2004 will illustrate the phenomenon. At Capel Curig, ten kilometres (six miles) east-south-east of Snowdon, 164 millimetres (6.48 inches) of rain fell in the 24 hours between 9 pm on Monday 2nd and 9 pm on Tuesday 3rd. During the three days of most intense rainfall, a total of 273 millimetres (10.75 inches) was collected between Monday 2nd and Wednesday 4th inclusive, while the six-day total from Friday 30 January to Wednesday 4 February amounted to 417 millimetres (16.43 inches). This is 25 per cent more rain than fell in the entire year in parts of Essex in 2003.

Capel Curig is certainly not the wettest place in Snowdonia. Hard under the eastern flank of Snowdon itself, places like Cwm Dyli, Llyn Llydaw, Glaslyn and Crib Goch routinely collect almost twice as much winter rainfall as Capel. Thus it is quite possible that at least 760 millimetres (30 inches) of rain fell in these locations during those six wet days, with as much as 300 millimetres (almost 12 inches) during the wettest 24 hours.

The orographically enhanced rainfall was not confined to Snowdonia. At Shap, in Cumbria, for instance, 185 millimetres (7.28 inches) fell over a five-day period, and at Eskdalemuir in Dumfriesshire 131 millimetres (5.16 inches) fell in three days. By contrast, to the lee of the Southern Uplands, only 4.5 millimetres (0.18 inch) fell at St Andrews in Fife; to the lee of the Lake District and northern Pennines just half a millimetre (0.02 inch) fell at Newcastle; and to the lee of the Welsh hills Birmingham recorded a mere five millimetres (0.2 inch).

Occasionally, if rarely, the atmosphere may become locked into a particular weather pattern for several weeks. When such a pattern favours heavy rain on windward slopes, some astonishing contrasts may be seen in the monthly rainfall figures. In March 1990, for instance, a total of over 1,000 millimetres (40 inches) fell on the hills above Glen Shiel in the western highlands, while in Aberdeen the month's aggregate was just 12 millimetres (0.48 inch). Glen Shiel's figure was four times the local average; Aberdeen's was less than a quarter of its normal amount. The effect can be seen with other wind directions too, though never to the same degree, as northerlies, easterlies and southerlies carry less moisture than westerlies and south-westerlies. Nevertheless, in February 1947, an exceptionally cold and snowy month dominated by easterly winds, the rainfall (actually, melted snow) on the fells above Upper Wharfedale in Yorkshire amounted to 165 millimetres (6.5 inches), compared with just 10 millimetres (0.4 inch) in Manchester. Further north, 125 millimetres (almost five inches) fell over the Aberdeenshire hills south of Aboyne, while the infamously wet rainfall stations at Glencoe and Glenquoich in the western highlands had no rain or snow at all during the entire month.

The contribution of this orographic effect can be seen in the long-term climate figures as well. Broadly speaking, the wettest places in the UK collect roughly eight times as much rain as the driest in an average year. Across the Scottish highlands, the average yearly rainfall ranges from 5,100 millimetres (just over 200 inches) on the slopes above Loch Quoich in Lochaber to 585 millimetres (23 inches) at Tarbet Ness in Easter Ross; in northern England the range is from 4,700 millimetres (185 inches) at The Stye, above Borrowdale, in the Lake District, to 535 millimetres (21 inches) at the mouth of the River Tees; and in Wales the annual rainfall at Glaslyn, high on the eastern flank of Snowdon, is 5,030 millimetres (198 inches), compared with 660 millimetres (26 inches) less than 50 kilometres (30 miles) away at Rhyl.

Where does rain go after it has fallen?

London's average annual rainfall is about 600 millimetres (24 inches). That represents about 7,100 tonnes of water per hectare, or 710,000 tonnes per square kilometre. (If you prefer, a yearly total of 24 inches of rain provides 2,500 tons of water per acre or 1.6 million tons per square mile.) The wettest parts of the Lake District, Snowdonia and Lochaber, where 4,000–5,000 millimetres of rain falls annually, collect almost six million tonnes of water per square kilometre in an average year. That's a heck of a lot of water. So where does it all go?

We all know that a short-term excess of rain can be disastrous: rivers bursting their banks, transport routes disrupted, and homes flooded. But during the course of a year, all that water disappears, taking its place in that endless journey which sustains life on our planet – the 'hydrological cycle'. Some of it finds its way directly into drainage channels, streams and rivers; this is called *runoff*. Some of it drains into the soil; this is *percolation*. Some vanishes directly into the atmosphere; this is termed *evaporation*. And some is absorbed by plants and from there transmitted to the air; this is known as *transpiration*. In the UK this complex process is remarkably efficient at redistributing those enormous quantities of water for at least 99 per cent of the time, the remaining 1 per cent constituting periods of flood and drought.

Given the great variability of British weather from week to week and year to year, there must be some sort of safety valve which enables the hydrological cycle to cope with run-of-the-mill downpours and dry periods. The role of safety valve is played by *groundwater* which is gradually used up (by evaporation and transpiration) during dry periods but which is regularly topped up (by percolation) when it rains. The same process applies on a regular annual cycle, which sees groundwater levels drop during the summer half of the year to be replenished during the winter half. In this way the water table rises and falls seasonally. In simple terms, then, the ground acts as a massive sponge which keeps drought and flood to a minimum.

It should not really be a surprise that in a climate like ours, which has fairly regular precipitation habits, the hydrological cycle works efficiently. The landscape has evolved over aeons to accommodate most of the meteorological vicissitudes we encounter in our particular corner of the planet. However, it will fail if the climate changes abruptly – at the beginning or end of an ice age, for instance – or if the natural environment is seriously tampered with. For example, building housing estates on the flood plains of major rivers is asking for trouble. And abstracting more water for industrial and residential use from the water-bearing strata than can be replenished naturally will also create serious problems. When a flood or a drought happens under these circumstances we have only ourselves to blame.

Evaporation

When we learnt about the British climate in our geography lessons we were taught that the UK lay in the temperate zone, on the western margin of a continent, where the prevailing westerly winds were responsible for mild rainy winters and cool moist summers.

Judging by the facial expressions of my classmates as well as the subsequent examination results, it was all in one ear and out the other for most of us. If, however, you were one of the few who absorbed those lessons, you will probably also remember that rainfall is fairly evenly distributed throughout the year in eastern and central England whereas there is a marked winter maximum further west, and that the hilly northern and western parts of Britain have most rain, whereas East Anglia has least. You might even recall the average yearly rainfall of your own neck of the woods: 600 millimetres (24 inches) in London, for instance, 700 millimetres (28 inches) in Edinburgh, 850 millimetres (34 inches) in Manchester, 1,000 millimetres (40 inches) in Cardiff, or 2,000 millimetres (80 inches) in Fort William.

But we were taught only half the story. At school I used to wonder why we were not allowed to play on the grass in February, which was supposed to be the driest month of the year. Why was the football field still muddy after three weeks without rain in winter, whereas in summer it was rock hard again just three days after a cloudburst?

Rain is no use at all in our gardens, or indeed in the reservoirs run by the water authorities to provide us with our domestic water supply, if it all evaporates before it does any good. Moisture evaporates from the ground when the air in contact with it is dry, and the process is accelerated by the warming effect of sunshine and assisted by strong winds which deliver a continuous supply of dry air. During an average July day in south Lancashire, for instance, 60 times as much moisture evaporates as on a typical January day.

In an average year in lowland Britain, evaporation exceeds rainfall during each month from April to August, and in parts of eastern England in March and September as well. Broadly speaking, that is why normal soils range from damp to waterlogged between late September and early March, but are dry for much of the time during late spring and summer. Evaporation is measured, by convention, in the same units as are used for rainfall, that is millimetres (or inches). In the Home Counties, and averaged over a 30-year period, monthly evaporation ranges from less than five millimetres (0.2 inch) in

December and January to just over 100 millimetres (four inches) in June and July; annual evaporation is 550–600 millimetres (22–24 inches). In west Wales, Cumbria, and the northern half of Scotland it is around 400 millimetres (16 inches) per year, and in the Western Isles little more than 300 millimetres (12 inches).

Over the annual cycle, therefore, in parts of south-east England and East Anglia evaporation slightly exceeds rainfall, and these areas could reasonably be called semi-arid. With a large population to support, domestic and industrial water supplies have to be imported from elsewhere, although the River Thames and the Great Ouse do a pretty good job of this.

In eastern and central England only the dullest and wettest – about 10 per cent – of summer months will have rainfall greater than evaporation. This is why systematic summer watering is necessary on farms and cricket grounds, in parks and gardens, in nine years out of ten in these parts of the country.

Warm seas mean heavy rain

The weather we experience at any one moment is the result of a complex interaction of season, time of day, geography, and the characteristics of the air-mass which happens to be in occupation at the time. These characteristics in turn depend on the area of origin of the air-mass, the speed at which it travels towards us, and the type of land or sea surface it has passed over on its journey. All these variables react one with another within a framework governed by the laws of physics.

One important aspect of season that is sometimes forgotten even by weather forecasters is the delay in the seasons over the waters which surround the British Isles compared with the dry land where we live. The range of temperature over the sea between summer and winter is also much smaller than it is over the land.

The main reason for both of these phenomena is the much greater thermal capacity of water compared with solid material such as rock and soil. Imagine placing a pan of water on a gas-stove. It takes many minutes until the water is too hot to touch. However, if you put an empty pan on the same stove the heated metal would burn your fingers within seconds. Thus in spring and summer, under the influence of strong sunshine, continents warm up rapidly and reach their hottest on average two or three weeks after the summer solstice, whereas the oceans warm very slowly and water temperature does not peak until about ten weeks after the solstice. The opposite pattern occurs during autumn and winter.

Thus British coastal waters are at their warmest in the first week of September when things are already cooling off over the land, and at their coldest in the first week of March when land temperatures are normally beginning to climb significantly. In the middle of the English Channel the range is from 8°C (46°F) to 17°C (63°F), off East Anglia it is from 5°C (41°F) to 16°C (61°F), beyond the tip of Cornwall from 9°C (48°F) to 16°C (61°F), and around northern Scotland from 7°C (45°F) to 12°C (54°F).

One important area where this seasonal delay has an important knock-on effect is in showery air-streams. Showers develop when there is a larger-than-normal temperature contrast between the lowest layer of the atmosphere and the air several kilometres above us. A showery south-westerly airflow affecting the UK during the autumn will produce fewer showers over the land as the season advances because the air in contact with the ground grows progressively cooler. The identical south-westerly flow over the English Channel remains comparatively warm and moist throughout the autumn, resulting in frequent heavy showers over the sea and along windward coasts. Indeed, autumn showers around the south coast are often so vigorous that they grow into thunderstorms. For the same reason, the showery activity usually dies down over the land at night, but it continues throughout the night at sea. A dramatic example of this mechanism occurred repeatedly during October and November 2000 across Sussex and Kent, exacerbating the already widespread floods which plagued those two counties during that rain-sodden autumn.

Why rain never turns to snow

The weather situation on 11 and 12 March 2004 could almost have been invented by a meteorology lecturer to illustrate snowfall in Britain. The weather chart on Thursday 11th was a classic: high pressure over Scandinavia feeding a very cold and dry easterly air-stream across the North European Plain towards England and Wales, a deep depression centred west of Ireland fuelling moist south-westerly winds heading our way from the relatively warm Atlantic Ocean, and an active occluded front marking the boundary between the two.

It is because such contrasting air-masses do not readily mix that fronts develop between them. The colder air-mass undercuts the warmer one, and the warm air is forced to rise, slantwise, over the cold air. This interaction results in the release of moisture in the form of clouds and rain (or snow), and that is why frontal systems are nearly always associated with bad weather.

Sure enough, the bad weather arrived, first of all in west Cornwall as rain and sleet on the morning of the 11th, then as snow over other parts of England and Wales as the front advanced northwards and eastwards. In many areas only trivial amounts settled on the ground, but by Friday morning there was a five- to ten-centimetre (two- to four-inch) cover from west Hertfordshire across to Gloucestershire, and ten to fifteen centimetres (four to six inches) in the Welsh Valleys, over the Brecon Beacons, and across the hillier parts of Herefordshire and south Shropshire.

Weather forecasters glibly talked of rain turning to snow as the weather system progressed across the country. This is not just sloppy, it is seriously misleading because rain never turns to snow; it is a physical impossibility. However, snow does turn to rain, all the time. In fact, much of the rain that falls in the UK starts off as snow, even in high summer, because the temperature at cloud level is frequently below 0°C. As the snow falls out of the bottom of the cloud, it travels through progressively warmer layers of air, so the snowflakes turn to raindrops.

So what happens when we perceive rain turning to snow? It is simply the result of the air beneath the cloud getting colder; in the March 2004 example it was a consequence of that strong easterly wind dragging chill continental air across the country. Thus, instead of the snow turning to rain as it fell, it just stayed as snow all the way down from cloud to ground.

April showers

April is associated with showers in our part of the world, not because they occur more often in this particular month compared with any other, but rather because the frequency of shower activity, inland at least, increases sharply between March and April.

Sunny mornings in winter usually presage fine days, but from April onwards sunny starts are sometimes deceptive, often being followed by a build-up of cloud which in turn leads to short sharp showers during the afternoons. Maybe it was this sort of weather that Shakespeare had in mind in *The Two Gentlemen of Verona*:

> O! how this spring of love resembleth
> The uncertain glory of an April day.

These springtime showers are a direct consequence of the growing strength of the sun at this season. Powerful morning sunshine warms the ground rapidly which in turn warms the lowest layers of the atmosphere. Bubbles of warm air – thermals – rise, gradually cooling again as they do so. Beyond a certain level, moisture condenses to form clouds, and, as the process continues, these cumulus clouds grow big enough and deep enough to produce showers.

Winds blowing from the west, north-west and north are the chief showery air-streams at this time of the year, and northerly winds are still cold enough to deliver showers of sleet, hail and snow rather than rain. Such snowfalls occasionally cover the ground, but normally do not last long.

Averaged over three centuries of rainfall recording in Britain, April has been the third driest month, after February and March, of the entire year. For short periods in the mid-eighteenth century, the late nineteenth century, and in the 1970s, it was the driest month of all. But since then we have moved into an era of very much wetter Aprils. In 1998, for instance, prolonged heavy rain over the Easter weekend led to serious flooding in the south Midlands in which six people lost their lives. April showers were almost a daily occurrence during April 2000, and were often extremely heavy, and frequently accompanied by hail and thunder. That was the wettest April, averaged over England and Wales, since 1756.

Chapter 2

Wind and warmth

Introduction

Along with moisture and motion, heat is one of the essential parts of the global weather machine. Surprisingly, perhaps, the air itself is not heated up by the sun to any important extent. Winter days of unbroken sunshine when the air is crisp and fresh and the temperature is below zero bear witness to that, as do sultry summer days when heavy clouds completely blot out the sun.

It is the Earth's surface that absorbs much of the heat energy that comes from the sun, and the lowest layer of the atmosphere is then warmed by contact with the sun-heated ground. This process is known as *conduction*. The warmth then spreads gradually outwards and upwards with the movement of the air thanks to a combination of processes, including *advection* (travelling with the prevailing wind), *convection* (rising and subsiding air currents triggered by temperature differences) and *eddy diffusion* (mixing of air by turbulence).

Not all the sun's energy that arrives at the top of the atmosphere gets through to ground level. Taking a long-term global average, just 7 per cent of solar energy is absorbed directly by the air itself; 16 per cent is absorbed by material suspended in the atmosphere, primarily water vapour, but also dust, salt particles, smoke, and so on; 25 per cent is absorbed by clouds or reflected by them back into space; and 4 per cent is reflected by the land, ocean and ice surfaces at ground level.

That leaves just under 50 per cent to warm up the Earth's surface. There are, of course, big variations from day to day, and from place to place, and on a cloudless day in the tropics as much as 80 per cent of the incoming solar energy may go towards heating up the ground.

Averaged over a year, the incoming and outgoing radiation balance each other. This is the weather machine's entire *raison d'être.* The travelling storms, the global wind systems and ocean gyres comprise the mechanisms whereby the excessive heat received at the tropics is redistributed to other parts of the planet. If they were not constantly there, busy doing their job, the tropics would get hotter and hotter, and the poles would get colder and colder.

This is a useful place to introduce so-called anthropogenic climate change – the warming of the planet caused by human activity. Human beings have been changing the climate ever since we started chopping down trees to build houses and to make fires: cutting down a forest changes the ability of that part of the Earth's surface to absorb radiation from the sun. Building cities, draining wetlands, reclaiming land from the sea, planting crops where once there was grassland, all result in tiny changes to the planet's radiation balance. Industrial processes, however, make changes on an entirely different scale. The emission of smoke and a variety of gases over the last 200 years or so has made significant changes to the radiation coming in and also to the radiation going out. The resulting changes in temperature have not all been in the same direction. Carbon-based gases such as methane and carbon dioxide have a warming effect, whereas nitrogen-based gases have a cooling effect, but the contribution by the warming gases is significantly greater than the contribution by the cooling gases.

Temperatures in the garden and underground

Why do people keep wine in a cellar? The answer is pretty obvious: it is sensible that good wine which is put down for perhaps several years should not be kept at a temperature which is either too high or too low, nor should it be subject to wide fluctuations of temperature between day and night or between summer and winter.

That sort of cool, steady-state environment is exactly what a decent cellar will provide, and the deeper the cellar is, the smaller the fluctuations in temperature are likely to be. The same is true of caves and explains why our distant ancestors who survived in the tundra-like climate which characterized western and central Europe during the last ice age chose to live in deep limestone caves – at Lascaux in south-western France for instance. That sort of environment may seem pretty cold and uncomfortable to us now, but when the outside temperature stood at –25°C (–13°F) as it often did during the winters of that era, a deep cave temperature of 4°C (39°F) would probably have seemed deliciously warm.

But have you ever stopped to ask why the temperature at great depth is constant? And have you ever wondered what sort of temperature fluctuations there are in the soil at smaller depths?

The greatest variations in temperature are found at the surface of the Earth, and are the result of the short-term imbalance between heat energy arriving from the sun (solar radiation) and heat energy escaping from the ground (terrestrial radiation). Solar radiation is greatest at midday on a clear day, when the sun is high in the sky, whereas terrestrial radiation is fairly constant for a given soil type. It is, however, most noticeable at night, when there is of course no solar radiation in the equation, and that is why the temperature falls at night especially when the sky is clear. A cloud cover will limit the incoming solar radiation, and it will reflect the terrestrial radiation back to the ground. That is why, broadly speaking, cloudy days are cooler than sunny days, and cloudy nights are warmer than clear nights.

An excess or deficit of heat energy at the surface of the ground is dissipated to the air above and to the soil below by several mechanisms. Conduction is the most important of these as far as the soil temperature is concerned. Soils comprise a mix of solid material, air and water, and some soils are better conductors than others. Dry sandy ones are relatively poor conductors of heat, whereas moist clayey ones are good conductors.

Thus sandy regions in the UK, for example, stabilized dunes in some coastal districts, the sandstone ridges of the south Midlands, the Weald of Kent, Surrey and Sussex, and the Breckland in south-west Norfolk, habitually experience a large range of temperature between day and night when the weather is calm and clear, with very high daytime extremes in summer, and very low night-time extremes in winter.

The temperature of the air in the shade has been recorded every day in this country for the best part of 250 years. Daily temperature recordings on the surface of the ground and underneath it do not have such a long history, but a handful of sites have continuous records of these elements extending back 100 years or more. These show us that the very wide fluctuations of temperature recorded on the surface of the ground become smaller and smaller as they are conducted downwards into the soil beneath. To illustrate this, let us look at some numbers.

On a sandy soil in the south-eastern quarter of England – near Thetford in Norfolk, for instance – the extremes of air temperature (recorded in a standard thermometer screen four feet above the ground) in a typical year are approximately 32°C (90°F) and –10°C (14°F). On a bare soil surface the range widens considerably with maxima as high as 55°C (131°F) and minima near –18°C (0°F).

A few centimetres below the surface of the ground the range of temperature quickly diminishes. The day–night and summer–winter fluctuations are still evident, but there is a delay in the time of maxima (and minima) because the conduction of heat downwards from the surface continues after the hour of maximum heating by the sun. Eight centimetres (three inches) below the surface the extreme temperatures are around 24°C (75°F) and –4°C (25°F), and 30 centimetres (one foot) down they are 21°C (70°F) and 0°C (32°F). Thus only in the very coldest winters does frost penetrate to a depth greater than 30 centimetres.

By the time we get 60 centimetres (two feet) below the ground surface the temperature variation between day and night is very small – less than 2 degC (3 degF) – the highest temperatures being in the early evening and the lowest around mid-morning. Once we reach 90 centimetres (three feet) – the precise depth varies with soil type – the day–night fluctuation disappears altogether. And at 1.2 metres (four feet) the highest reading, typically 17°C (63°F), occurs in early September, while the lowest reading, about 3°C (37°F), is usually obtained in early March. Day-to-day temperature changes at this depth are always very small, measured in fractions of a degree.

Measurements are not normally taken at greater depths, but a series of observations were made by a Colonel H. S. Knight at Harestock, near Winchester, in the 1890s. He sank a number of boreholes of varying depths in his grounds and hung thermometers in them on lengths of chain. The thermometers were specially constructed, with the bulbs encased in paraffin wax. The wax acted as an insulator, so that when he recovered each of the instruments in order to make his observation, the mercury reading did not change.

His records showed that at three metres (10 feet) the temperature ranged from 8.2°C (46.8°F) in April to 11.7°C (53.1°F) in October, and at six metres (20 feet) the warmest month was December with 10.4°C (50.7°F) and the coldest was July with 9.5°C (49.1°F). So at this depth the normal annual cycle in the air above was completely reversed, indicating that it took six months for temperature changes in the atmosphere to take effect six metres below ground.

At nine metres (30 feet) Colonel Knight found that there was no evident annual cycle at all, and all months fell between 9.6°C (49.3°F) and 10.2°C (50.4°F), while at a depth of just over 20 metres (70 feet), his deepest borehole, the temperature appeared to be constant at exactly 10.0°C (50.0°F). These nineteenth-century experiments suggest a twenty-first-century experiment for those who are not convinced that our climate is getting warmer. Sink a borehole 20 or 30 metres deep, suspend a very sensitive temperature sensor in it, and take regular readings. If the atmosphere is getting warmer, the deep soil temperature will also climb, but without the wide daily oscillations that tend to obscure any slow underlying trend in air temperature.

The urban heat island

When I first came across the expression ‘London’s urban fabric’ in my youthful meteorological reading, I was instantly presented with a bizarre mental image of some omnipotent being reupholstering the capital’s skyline with ginghams and chintzes. Sadly, the context was rather more mundane.

The urban fabric, that is, the patchwork of asphalt, brick houses, slates and tiles, gardens, concrete tower blocks, parks and playing fields, is crucial to our understanding of how our towns and cities modify the climate in their immediate vicinity. Up to a point, they create a distinct urban climate, perceptibly different from that of the surrounding countryside. This effect is known as the ‘urban heat island’ because the most important climatic consequence of urbanization is an increase in temperature. There are other effects, including a marginal increase in rainfall, a decrease in average wind speed but an increase in gustiness, a slight drop in humidity levels, and a drop in frequency of frost and snow. Fog may become more frequent if the city is afflicted by atmospheric pollution containing particulate matter (London before the 1956 Clean Air Act, for instance), or less frequent if the main influences are simply higher temperature and lower humidity (such as London after the clean air legislation was enacted).

Some commentators who seek to dismiss the warming trend in our climate as of little consequence often refer to the warmth of urban areas as an explanation. Most weather-recording sites, they say, are now in towns and cities whereas a hundred years ago they were in the countryside, and climatologists are such poor scientists, they imply, that they have yet to realize this. I hate to disillusion the poor lambs, but the concept of the urban heat island was understood long ago. One of Britain's meteorological pioneers, Luke Howard, wrote in 1818 about London:

> . . . the temperature of the *city* is not to be considered as that of the *climate*: it partakes too much of artificial warmth induced by its structure, by a crowded population, and the consumption of great quantities of fuel.

Moreover – it may sound paradoxical but it is true – many meteorological observatories in Victorian times were located in cities or on their fringes, often associated with universities, learned societies, municipal authorities or industrial concerns, whereas present-day recording sites are predominantly rural, at agricultural colleges, on forestry commission land or adjacent to airfields. In any case, where long-standing records have been kept in places which were once rural but are now suburban, climate experts have long since (that is, since at least the 1940s) analysed the effect of urbanization by reference to nearby sites which have been continuously rural or continuously urban. The warmth of towns and cities has, therefore, been accurately allowed for in all detailed studies of climate change.

In the UK, the urban heat island effect is most prominent on calm clear nights when a city centre may be as much as 12 degC (22 degF) warmer than the surrounding countryside, but it diminishes rapidly as soon as the wind begins to blow, and it vanishes altogether when the wind becomes strong.

The city of Bath was the site of the first systematic study of the climate of an urban area. The study was undertaken by two geographers, William Balchin and Norman Pye, and was published in 1947. Over the next two decades many similar analyses were carried out in towns and cities around the country. Probably the biggest and most detailed of these was conducted by a team led by Tony Chandler in London between 1958 and 1964.

Although Luke Howard wrote about it in his *Climate of London* almost two centuries ago, the paper presented by Balchin and Pye was the first to quantify the changes.

The materials from which towns and cities are constructed are very different from the grass, trees and crops of the surrounding countryside. The irregularity of the city surface also inhibits the movement of air, and when there is insufficient turbulence for the wind to penetrate to ground level the air becomes stagnant in the streets and squares. Most building materials are good at absorbing heat energy from the sun, and the darker the colour of the material the more solar radiation is taken in and the hotter it becomes. Place the flat of your hand on the roof of a black car on a sunny day and you will find it much hotter than the roof of a white car. Thus tarmac becomes particularly hot on clear summer days. The heat absorbed during the day is gradually released at night, rather like a giant night-storage heater, and that is why it can be so difficult to sleep in inner-city districts during hot weather: the August 2003 heatwave was a case in point, with midnight temperatures in central London around 26 or 27°C (around 80°F), dropping no lower than 21 or 22°C (70–72°F) by dawn. It was even worse in Paris where midnight readings of 30–32°C (86–90°F) were typical, with nothing lower than 25°C (77°F) at daybreak.

However, any breeze will help disperse pockets of abnormally warm or cold air, and with a sustained wind of 13 knots (15 mph) the warming effect of urban areas is almost negligible, disappearing completely at 18 knots (21 mph) or more. But when there is no air movement at all, the so-called 'urban heat island' can be very pronounced especially at night. The biggest temperature contrasts are most likely to be found just before sunrise after a clear, calm night, and the contrast will be accentuated if the ground is dry or snow-covered. One of the most extreme examples of this occurred on the morning of 14 January 1982, a few days after a severe snowstorm, when the minimum temperature in central London was –5°C (23°F), while at several sites in the Hertfordshire and Buckinghamshire countryside it was –21°C (–6°F).

Increased rainfall is more likely to occur in large conurbations in tropical regions, while in cool temperate latitudes like ours the evidence is equivocal. There are comparatively few occasions during the year when the higher temperature of the air in British cities is sufficient to trigger those rising currents of air – thermals, or convection currents – which in turn produce shower clouds and thunder clouds. Even when they do occur, the wind at cloud level is normally sufficient to carry the clouds beyond the city boundary before any rain actually falls. A study of London's rainfall in the 1980s catalogued only 11 days with measurable rain in central London when none fell at a selection of suburban and rural sites.

During the same period there were 31 occasions with no rain in central London when measurable rain fell at a minimum of six of the seven peripheral weather stations.

Blocking highs and anticyclonic gloom

Westerly weather is typical of the majority of British winters, especially in recent years. The frequency of winds from the Atlantic increased sharply from 1988 onwards, and the strength of these westerlies increased threefold between the 1960s and the 1990s. Occasionally, though, the broad-scale weather pattern over Europe and the northern part of the Atlantic Ocean undergoes a major reorganization, and sometimes this results in the replacement of the normal 'Icelandic low' with a high-pressure system. Such a system is known as a 'blocking high', so called because it blocks the normal west-to-east progress of Atlantic weather systems across the British Isles and Europe.

During westerly episodes in winter, Britain's weather is often very disturbed, with frequent and rapid changes: prolonged downpours accompanied by rough winds alternating with days of bright but blustery weather. The weather is usually worst in north-western Britain, especially where high ground faces into the wind, and in the highlands and islands in particular these periods are often wild, windswept and thoroughly rain-soaked. For these areas especially, the arrival of a blocking high is good news, for as long as the high-pressure system lasts, the rain and the wind will be held at bay.

High-pressure systems – 'anticyclones' in meteorologists' jargon – are supposed to bring fine and settled weather while low-pressure systems – depressions – bring rain. At least that is what we were taught in geography lessons at school. However, not all anticyclones bring sunshine, and this is particularly true during the winter half of the year.

The sort of weather we get during a 'blocked' episode in winter depends on precisely where the blocking high takes up residence. If it sits bang slap over the UK it may bring widespread frost and patchy fog at night, but some sunshine by day, at least to start with. If it settles over Scandinavia or the northern North Sea it will feed bitterly cold easterly winds from the continent, with the temperature close to zero all day and a threat of snow in eastern and southern parts of Britain. If it is located near Iceland it will open the back door to a northerly airflow plunging across the British Isles directly from the Arctic.

The most usual location, however, is over the Atlantic to the west of Ireland or Scotland, but to the south of Iceland. The spell may begin with a short-lived Arctic plunge, bringing bright sunshine, snow showers and night frosts, but the supply of cold air is soon cut off as moist and somewhat milder Atlantic air travels around the northern flank of the high, thence turning south-eastwards towards the British Isles. Although the wind may still blow from the north or north-west, its origin is not a particularly cold one; as a consequence, our weather is often grey and damp with persistently overcast skies, occasional drizzle and poor visibility, but the temperature remains close to the seasonal average.

When a winter anticyclone first establishes itself over the British Isles, most regions enjoy a couple of days of blue sky and bright sunshine although it is usually very frosty at night and patchy fog readily forms. After a couple of days, moist air from the Atlantic Ocean usually penetrates the anticyclone's circulation, forming a layer of stratocumulus cloud about a kilometre (half a mile) above the ground, and the winds aloft which blow sluggishly in a clockwise manner around a 'high' carry this cloud sheet to all parts of the country.

When this happens in summer the cloud layer is usually broken up by turbulent air motion – convection currents, to be accurate – caused when the lowest layers of the atmosphere are warmed up during the daytime even when the weather is cloudy. In late autumn and winter the sun is comparatively weak and nights are long, and the layer of air near the ground quickly becomes colder than the air above it. When the wind blows, this cold layer is soon dispersed, but under an anticyclone where there is little or no wind it is often strongly developed. This reversal of the usual fall of temperature with height above the ground is called a 'temperature inversion' and it acts as a sort of atmospheric lid, trapping pollution, cloud, mist and moisture in the cool layer of air near the ground. Meteorologists call this sort of weather 'anticyclonic gloom'.

High-pressure systems can be difficult to shift, and a blocking high may persist over or close to north-west Europe for two or three weeks at a time. Periods of anticyclonic gloom can therefore be remarkably persistent and – ironically – very depressing. Nevertheless there are often subtle day-to-day changes in the position of the high, and of the gentle winds in its circulation, and temporary incursions of slightly drier and cooler air can bring a sudden, if short-lived, improvement in the weather.

The first indications of the end of a blocking episode come when the centre of high pressure drifts gently eastwards into the continent. Here it may remain for several more days before it finally collapses. Recently a new nickname has come to prominence for a high centred to the east or south-east of the UK: the 'Eurotrash high' – neatly highlighting the pollution, fog and low cloud which drift off the continent across England when such an anticyclone stagnates over Germany or central Europe.

The jet-stream

Although the weather has been called 'the great British obsession', you only have to compare the coverage in newspapers and on television in the UK and the USA to appreciate that the Americans are even more obsessed with it than we are. They have had a weather channel on cable television across the pond for many years, but all efforts to get an equivalent channel off the ground in Britain and/or Europe – and there have been several attempts in the last ten years or so – have been dismal financial failures.

There are several differences in style and content between British and American television weather forecasts. Perhaps the most notable of these is the use of maps showing the changes in position and strength of the jet-stream. These have been central to TV weather forecasting in the US for decades, but conspicuous by their absence over here. Francis Wilson certainly made an effort to emulate his transatlantic counterparts during his days in the 1980s as the main weather presenter on the breakfast-time programme on BBC1; also in the 1980s I used a large chart showing the jet-stream position over the Atlantic and Europe to illustrate the forecast for the week ahead in the sadly short-lived *Sunday Today* newspaper. All other television and newspaper outlets steered well clear of the subject, and continue to do so to this day.

Jet-stream maps may be regarded as a gimmick by some people, but that would be missing the point. The mid-latitude jet-stream is crucial to the understanding of the travelling weather systems which affect such a large part of both the northern and southern hemispheres. It exercises such tight control on these weather systems that an analysis of its behaviour is a vital first step to predicting what the weather is going to do to us during the coming week. Superimpose the position of the jet-stream on one of those satellite images which show the distribution of cloud across, say, Europe and the Atlantic, and the intimate relationship between the two will immediately be apparent.

So what is the mid-latitude jet-stream? It is a concentrated zone of rapidly flowing wind high in the sky, usually found near the boundary between the troposphere and the stratosphere – somewhere between eight and thirteen kilometres (five and eight miles) above the Earth's surface. It may be only a few hundred kilometres across but it can be several thousands of kilometres long, rather like an enormous atmospheric conveyor belt, and at its core the wind blows at a sustained speed of 125 knots (145 mph) and sometimes at 220 knots (250 mph) or more. These jet-streams almost encircle the hemisphere, generally flowing from west to east; long-haul passenger aircraft often catch a ride on one, resulting in early arrivals at Heathrow from North American points of departure.

The strongest part of the jet-stream is associated with the rapid development of depressions on the sea-level weather charts. A relatively innocuous disturbance lying beneath the right-hand side of the entrance (in the northern hemisphere) to the core of the jet-stream will then travel the length of the conveyor belt, finally leaving beneath the left-hand side of the jet's exit, by now a large and powerful mid-latitude storm. For this reason both the right-entrance and the left-exit are known to meteorologists as 'development areas'.

Mid-latitude depressions

When seen from space, Earth appears as a bright, blue-green disc with white swirls and streaks draped across it. It may look rather pretty and peaceful from such a great distance, but those decorative patterns highlight the areas where atmospheric processes are at their most turbulent. The largest of those white swirls are actually mid-latitude depressions, and they stalk the temperate regions of both hemispheres during all seasons of the year. Satellite images show that in the northern hemisphere they chase each other across the Pacific and Atlantic Oceans at irregular intervals, and in the southern hemisphere they race frantically around the Southern Ocean with scarcely a break.

The biggest measure more than 3,000 kilometres (almost 2,000 miles) across and may take several days to pass over a particular point on the planet's surface. They are far too large for the detail of how they work to be detected by a ground-based observer, but they are clearly described on weather charts, and they are often magnificently displayed on a time-lapse sequence of satellite pictures. Their movement and development are, as we have noted, controlled by the mid-latitude jet-streams.

Even before the advent of satellites, scientists had learnt enough about these depressions to have devised a good 'model' of their workings. This model was formulated by three Norwegian meteorologists – Vilhelm Bjerknes, his son Jakob, and Halvor Solberg – during the 1910s at the Bergen School of Meteorology, which was one of the first university meteorology departments in the world. They examined every Atlantic depression that approached the Norwegian coast over a period of several years, and they discovered that the sequence of events fell into a recognizable pattern for the great majority of these disturbances.

The typical depression starts with two contrasting air-masses flowing side by side. One is of subtropical origin and is warm and moist; the other is of polar origin and is very much colder and clearer. The boundary between these two masses of air is often very sharply defined, bringing abrupt changes in temperature, wind and weather as it passes over any particular place. In wartime Europe it was natural to liken the air-masses to two huge armies advancing and retreating, and the boundaries between them to battle fronts. The Bjerkneses named the atmospheric battle lines between the two air-masses the 'polar front'. Sections of the polar front marking the advance of warm air became 'warm fronts', and those where cold air was moving ahead became 'cold fronts'.

Sometimes a kink or wave forms on the polar front. On the forward side of the wave the warm air advances, creating a warm front, while behind the wave the cold air cuts deep into the rear flank of the warm sector, creating a cold front. The barometric pressure drops rapidly at the point of the wave as air is drawn upwards and outwards by the jet-stream aloft. As the pressure drops, a centre of low pressure forms, and the wind now begins to blow around this central point, in an anticlockwise sense in the northern hemisphere and clockwise in the southern. As the depression develops further, the pressure at the tip of the wave drops lower and lower, the winds encircling it grow stronger, and the depression is said to be 'deepening'. The wave becomes narrower and narrower as the cold front gradually catches up the warm front, the slice of warm air between being progressively forced upwards, slantwise, above the two fronts. The jet-stream high above drives the whole system forwards from west to east.

No two depressions are exactly alike, but an observer on the ground will notice the same approximate sequence of events, as the slanting warm front first appears high in the sky, then approaches closer and lower, eventually passing over the observer's location to be followed by a period in the tropical air – the warm sector – before the cold front arrives and the cold air kicks in again as the depression moves away.

It says much for the original Bergen model that it remains a fundamental feature of mid-latitude weather charts to this day, even though it was devised when meteorologists knew little about what was happening in the upper atmosphere. More recent research using all sorts of data gathered from various levels of the troposphere including many different kinds of satellite image has enabled modern meteorologists to make a number of refinements, but the basic accuracy of the Norwegian meteorologists' analysis is still accepted.

Secondary depressions

Severe gales occur most frequently around Britain's western and northern seaboards, and the Hebrides, Orkney, Shetland and the Western Highlands are among the windiest regions of Europe. But short-lived gales, maybe lasting six hours or so, can develop suddenly in any part of the United Kingdom and at any time of the year although they are most frequent in late autumn, winter and early spring.

These short spells of wild weather are brought to us by what meteorologists call 'secondary depressions', or simply 'secondaries'. In some ways this is a misleading name since it implies such a creature is less of a threat than a 'primary depression', but in reality this is not so. Primary depressions are relatively old, mature ones, and like mature people they tend to become a bit flabby in the middle, rather sluggish and no longer as energetic as they once were.

A secondary depression forms in the outer circulation of the original storm and feeds off its energy. The primary could thus be regarded as (and is often called by weather forecasters) the 'parent depression', whereas the secondary is its offspring. And just like human offspring, secondary depressions are often vigorous, quickly growing, rapidly moving and can also be prone to violence. Some, of course, are more ferocious than others, and the strength and orientation of the jet-stream, as we have seen, plays a crucial part in deciding which ones develop dramatically, and which ones do not.

Two noteworthy destructive autumn storms, one of them very famous, the other not as well known, were caused by vicious secondaries. The Great Storm of October 1987 was the famous one, but another, nearly as destructive, zoomed in off the Atlantic Ocean in October 1964. There could hardly have been a larger contrast in coverage between these two storms, both in the popular media and in the relevant meteorological journals. The 1987 event was, of course, analysed in minute detail, whereas the 1964 gale was almost completely ignored.

The reason for the contrast is that the secondary depression of 9 October 1964 tracked along the length of the English Channel, and, with strongest winds confined to its southern flank, the British mainland escaped. Not so the Channel Islands, where the gale was more intense even than the 1987 one. Between 2 pm and 3 pm that day the westerly wind averaged 68 knots (78 mph) with a peak gust at 2.25 pm of 94 knots (108 mph). Hundreds of hectares of glasshouses on both Guernsey and Jersey were destroyed by these ferocious winds, and the late tomato crop was completely ruined.

Probably the most dramatic secondary depression in recent times swept across central Ireland and southern Scotland during the night of 7–8 January 2005. The winds were strongest in a broad zone across Ireland and northern England, with peak gusts at low-level sites of 70–85 knots (roughly 80–100 mph), and at St Bees Head on the Cumbrian coast a gust of 91 knots (104 mph) was recorded. Heavy rain had already been falling for two days, but a renewed downpour associated with the secondary brought 72-hour rainfall totals over the Lake District to 200–300 millimetres (eight to twelve inches). Floodwaters rose rapidly along the length of the River Eden, and the flooding in Carlisle was very serious indeed, described locally as the worst for over 50 years.

Chapter 3

The four seasons: spring

Introduction

The passing of a century has given us a valuable opportunity to take stock of our changing climate. The British climate has never been static, although the experts are now very concerned about the extent to which human activity is contributing to the present warming trend. That debate is often carried into the popular media with scant reference to facts and figures; let us, for a change, look at the records and the statistics without feeling the need to draw trite conclusions from them.

Spring is defined differently in different disciplines. The astronomical definition, often found in almanacs and diaries, has it starting at the spring equinox and ending at the summer solstice – approximately from 21 March to 21 June, with small variations in the dates resulting from our position in the leap year cycle. In the medieval Church calendar it ran from 22 February to 24 May – *dat Petrus ver Cathedratus* – and while Britain was a largely rural society, spring was popularly regarded as comprising February, March and April. Nowadays it is more usually March, April and May, and this equates to the spring quarter as used in climatological account keeping.

It is important to remember that these three-month periods were selected for calculating seasonal climate statistics in an era when observations were tabulated by hand and averages were calculated in the head. Using three calendar months was simply a book-keeping convenience.

Professor Hubert Lamb, one of the greatest figures in twentieth-century climatology, analysed seasonal variations in the weather of north-west Europe, and found that there were specific times when major changes in weather patterns took place. Between these dates the climate appeared to be relatively stable and was characterized by lengthy spells of one kind of weather or another. He called these the *natural seasons* and there are five of them, although none equates exactly to spring. Lamb's seasonal division includes *late winter and early spring* which runs from about 20 January to 31 March, and *spring and early summer* which lasts from approximately 1 April to 17 June.

Sadly, the British climate is such a variable one that the seasonal progression is usually very different from one year to the next, and even Professor Lamb's analysis fails to cope with these variations in many years. It is also difficult to see a five-season year catching the public's imagination. So for better or worse we appear to be stuck with the simple division of the year into four quarters of three calendar months each.

March in the twentieth century

> The flowers that bloom in the spring,
> Tra la,
> Breathe promise of merry sunshine;
> As we merrily dance and we sing,
> Tra la,
> We welcome the hope that they bring,
> Tra la,
> Of a summer of roses and wine.
> W. S. Gilbert (*The Gondoliers*)

A scientist would argue that Gilbert got it the wrong way round: it is the warmth of spring sunshine that brings forth the flowers that cheer us up during March, April and May. But then po-faced scientists never figured prominently in the topsy-turvy world of G and S.

In practice, the true beginning of spring can vary enormously from one year to the next. In some years, 1998 for instance, plant growth scarcely stops at all during the winter, and the daffodils are out as far north as Glasgow and Edinburgh before the end of February. In others, most recently in 1996, growth is so sluggish that the countryside forgoes its spring garb until well into April.

Very roughly, significant plant growth only occurs when the temperature is above 6°C (43°F), and that is close to the long-term average temperature for the month of March over a large portion of the UK. There will, of course, be variations either side of this 'normal' value, both from place to place and from year to year. If we arbitrarily select that 6°C figure as separating winter months from spring months, the following interesting facts emerge. In the comparatively mild West Country, Plymouth (average March temperature 7.4°C or 45.3°F) has had 90 of the last 100 Marches in the 'spring' category, while in the relatively cold east of Scotland, Edinburgh (average March temperature 5.1°C or 41.2°F) has had only 25 Marches above the threshold while the other 75 Marches were 'winter' months.

We can use statistics representative of the whole of England and Wales to look back at March over the last 120 years or so, decade by decade, to see how it has changed since our great-grandparents' days (see page 40).

The first two decades of the twentieth century were the coldest, followed by a marked warming in the 1920s, but then there was a slow slide backwards to another cold episode during the 1960s, 1970s and early 1980s. The century ended with a dramatic upward surge of warmth, mirroring the large rises in temperature during the same years in both January

Decade	*Mean temperature*	*Rainfall*	*Sunshine*	*Snow cover*
1880–89	4.7°C	65 millimetres	107 hours	–
1890–99	5.4°C	58 millimetres	117 hours	1.8 days
1900–09	5.2°C	72 millimetres	107 hours	1.5 days
1910–19	5.2°C	80 millimetres	96 hours	2.2 days
1920–29	6.1°C	53 millimetres	122 hours	1.1 days
1930–39	5.8°C	55 millimetres	124 hours	2.1 days
1940–49	5.8°C	60 millimetres	110 hours	3.7 days
1950–59	5.9°C	62 millimetres	109 hours	1.9 days
1960–69	5.6°C	62 millimetres	109 hours	1.7 days
1970–79	5.5°C	69 millimetres	106 hours	2.3 days
1980–89	5.8°C	94 millimetres	93 hours	1.0 days
1990–99	7.3°C	60 millimetres	114 hours	0.4 days
2000–05	7.0°C	62 millimetres	111 hours	0.7 days

and February, but that warming process appears to have stalled since 2000. Applying the 6°C (43°F) threshold in order to separate 'winter' months from 'spring' months, between 1900 and 1987 Marches divided roughly fifty-fifty; but from 1988 onwards there have been 15 'spring' Marches and only three 'winter' Marches.

On average, March is 2–3 degC (roughly 3–5 degF) colder than April, but the Marches of the 1990s have been very nearly as warm as the Aprils of the 1970s; digging deeper into the record books we find that the Aprils of the 1880s (and several earlier decades) were actually colder than most recent Marches. As one might expect from the advancing warmth, March snowfalls have almost become a thing of the past.

The first 20 Marches of the twentieth century were not only cold, they were also wet and gloomy, but the 1920s and 1930s were characterized by Marches which were mostly warm, dry and sunny. There followed a prolonged decline which culminated in a remarkably wet and sunless decade in the 1980s, but this deterioration was abruptly halted in 1989. The five coldest decades averaged 76 millimetres (2.99 inches) of rain and 102 hours' sunshine, whereas the five warmest decades averaged 58 millimetres (2.28 inches) of rain and 115 hours' sunshine. This association of warmth with sunshine and lack of rainfall is a characteristic of the summer half-year, in marked contrast to the late-autumn and winter months when the coldest decades were dry and the warmest ones wet.

During the twentieth century, March brought us some astonishing extremes of weather. As befits a transitional month, it produced wintry spells which matched anything January and February threw at us, as well as sudden early heatwaves which would not have disgraced June or July.

March disaster of the century

One single March, that of 1947, was so outrageously bad that it qualified for biggest freeze-up, worst blizzard, worst flood and worst gale. It marked the closing stages of that infamous winter, and much of the country was snowbound for the first half of the month. Power cuts occurred daily, and the image of Buckingham Palace and the Houses of Parliament working by candlelight less than two years after victory was won in war was a potent one. A sequence of snowstorms swept the country, paralysing transport. By the 6th, level snow lay 38 centimetres (16 inches) deep in Birmingham, and 150 cm (five feet) deep at Clawdd-newydd, near Ruthin, in north-east Wales. A rapid thaw set in in mid-month leading to the most widespread inland floods on record in Britain, and then on the 16th a violent gale swept southern and central England, causing extensive damage as sustained winds exceeded 45 knots (52 mph), and gusts up to 85 knots (98 mph) were logged at Mildenhall in Suffolk.

Heatwave of the century

March 1965 began in the freezer and ended in the oven. A record-breaking heatwave began on the 27th, and two days later the mercury soared to 25°C (77°F) in Norfolk and Yorkshire. The fine spell lasted over a week, but apart from a couple of days in mid-May the temperature in some places never again reached such dizzy heights that year, and at Whitby on the north Yorkshire coast 29 March was actually the hottest day of the entire year.

Drought of the century

Dry weather in late winter and spring is feared by farmers and water companies alike because if the land is drying out before summer gets under way there is a strong threat of a damaging drought by the end of the summer – especially if it is a hot, dry season. Such a sequence of events occurred in 1938: several places in southern England recorded no rainfall at all during March and April, and by August standpipes had been erected in several towns and cities to provide water to householders when domestic supplies were curtailed.

April in the twentieth century

'Daffy snow,' my old granny used to call it. Those sudden, sometimes heavy, falls of snow which turned up unexpectedly in late March or early April would flatten the massed ranks of daffodils in the garden, only for them to recover completely as the snow melted the next day. And later in the season, towards the end of April or perhaps even during the first week of May, a sudden wintry shower would bring soft, fat snowflakes, mingling with the swirling clouds of cherry blossom swept from the branches by a boisterous north wind.

These unwelcome but usually brief reminders of winter seem to be a thing of the past. We had a few passing snow flurries in early April 2003, but they served merely to emphasize the absence of 'daffy snow' throughout the last decade and a half. Once again the warming trend in our climate during the last hundred years or so has had a knock-on effect in our gardens and in the countryside. Recent Aprils have delivered not just fewer snowfalls but fewer damaging frosts too, while the bulbs, the leaf-buds and the blossom all arrive two or three weeks earlier than they used to.

A statistical analysis of April since 1880 will help us to put the recent warming into some sort of historical context. The Januarys, Februarys and Marches of the 1990s were all warmer than anything that had gone before, but the same is not true of April – not quite, anyway. These figures are averages representative of England and Wales as a whole:

Decade	*Mean temperature*	*Rainfall*	*Sunshine*	*Snow cover*
1880–89	7.3°C	59 millimetres	147 hours	–
1890–99	8.1°C	59 millimetres	146 hours	0.4 days
1900–09	7.6°C	59 millimetres	164 hours	0.7 days
1910–19	7.7°C	53 millimetres	153 hours	1.2 days
1920–29	7.6°C	65 millimetres	137 hours	0.2 days
1930–39	8.0°C	67 millimetres	121 hours	0.1 days
1940–49	9.3°C	52 millimetres	167 hours	0.1 days
1950–59	8.1°C	49 millimetres	153 hours	0.2 days
1960–69	8.2°C	71 millimetres	126 hours	0.2 days
1970–79	7.6°C	56 millimetres	124 hours	0.5 days
1980–89	7.9°C	58 millimetres	143 hours	0.5 days
1990–99	8.6°C	68 millimetres	150 hours	0.1 days
2000–04	8.8°C	84 millimetres	153 hours	0.2 days

The statistics show that the first 30 Aprils of the twentieth century were consistently cold, but this period was followed by a dramatic warming during the 1940s and early 1950s when Aprils averaged almost 2 degC (3 degF) warmer than they had been in the 1920s. Cold Aprils returned from the late

1950s onwards, culminating in some of the coldest months of the century in the 1970s. The sharp warming of the 1990s, characteristic of practically every month of the year, shows up as well, but the giddy heights of the 1940s have not yet been emulated. Surprisingly, there is no clear association between April temperatures on the one hand, and April rainfall or sunshine on the other. The driest decade was the 1950s and the wettest was the 1960s, and both had roughly the same mean temperature. However, the exceptionally warm 1940s were fairly dry and mostly sunny. The 1930s, 1960s and 1970s all had a marked shortage of sunshine, and the 1930s threw up the astonishing statistic that April was during that decade actually a duller month than March.

Over the years April has produced some extraordinary individual weather events which do not show up in the statistical summaries. Here are a few of them:

Snowstorm of the century

It was either 1908 or 1981 – take your pick. Both storms came during the fourth week of April, on both occasions just a few days after the Easter holiday. Worst hit in 1908 were Hampshire, Berkshire, Oxfordshire and parts of neighbouring counties, and in the triangle between Andover, Newbury and Swindon level snow lay 60–75 centimetres (24–30 inches) deep on the morning of the 26th. At Oxford, the 45-centimetre (18-inch) fall was deeper than any winter snowfall here during the entire twentieth century although the light winds meant that there was precious little drifting in the city. By contrast the 1981 snowstorm was piled into huge drifts by a northerly gale, and worst hit were hilly regions from the Yorkshire Dales and Pennine Moors in the north to the Mendips and Salisbury Plain in the south. Undrifted snow lay 60 centimetres (24 inches) deep on the fringes of the Peak District in Derbyshire, power and telephone lines were brought down, and hundreds of spring lambs were lost.

Flood of the century

Less than a decade ago a prolonged downpour during the Wednesday and Thursday before Easter deposited 50–80 millimetres (2 to 3 inches) of rain – six weeks' worth – in a broad band across the south Midlands. It was 1998. Almost 125 millimetres (5 inches) fell just north of Banbury. The water rapidly found its way into the catchments of the Cherwell, Avon and Nene rivers, and devastating floods hit Leamington, Northampton, Peterborough and Wisbech. Six people died, and a government enquiry was called to discover why the warnings of flooding were inadequate. A new flood warning service, under the auspices of the reorganized Environment Agency, was set up within two years – just in time for the autumn and winter floods of 2000–01.

Heatwave of the century

Another Easter event, this time in 1949, saw southerly winds draw exceptionally warm air from subtropical latitudes of the Atlantic Ocean, somewhere near the Canary Islands, northwards across Spain and France towards the UK. Saturday the 16th was the hottest day of this spell, and by mid-afternoon the temperature had climbed above 27°C (80°F) over much of England, peaking at 29.4°C (85°F) at Camden Square in London. This remains the highest temperature ever recorded on any date in April in the British Isles.

Curiosity of the century

April is the peak month for 'Fen Blows'. These occur when gales coincide with a protracted dry spell. The light soils of the Fens, not yet secured by growing crops, are lifted by the strong winds to a height of several hundred metres, stripping fields, ruining young shoots, and filling drainage dykes with silt. One of the worst of these happened in late April 1955 in and around the Cambridgeshire village of Manea. The incidence of fen blows has diminished in the last two decades, thanks to the reintroduction of hedges to act as windbreaks, and a return to agricultural fashion of winter sowing.

May in the twentieth century

> Rough winds do shake the darling buds of May,
> And summer's lease hath all too short a date . . .

Shakespeare was comparing the intended recipient of the sonnet to an English early summer's day, and found the summer weather wanting. Damned with faint praise, a sensitive soul might think, given the fickleness of a typical May. Oddly, rough winds are not normally a feature of May's weather; severe gales are rarer in this month than at any other time of the year, although brisk breezes are common enough. Had the Bard been climatologically correct, he might have worked in something about unexpected cold snaps or late frosts, but somehow one feels that it would not have sounded quite as good.

Given its propensity for Arctic winds and sharp frosts, May can scarcely be considered a summer month. True, the sun is as high in the heavens as it is in late July and early August, and, given a cloudless sky and no wind, it can surely shine strong and hot. But on most May days the air is cool and the breeze is fresh. Temperatures over 20°C (68°F) do occur from time to time, but on some days the mercury sticks resolutely around 7 or 8°C (45–47°F) and keen overnight frosts can cause much damage to young plants.

Any underlying trends in Britain's May climate during the last 125 years should be highlighted in the statistical summary of the month. These decade-by-decade averages are representative of England and Wales:

Decade	*Mean temperature*	*Rainfall*	*Sunshine*	*Thunderstorms*
1880–89	10.9°C	62 millimetres	189 hours	–
1890–99	10.9°C	55 millimetres	198 hours	2.0 days
1900–09	10.8°C	62 millimetres	213 hours	2.4 days
1910–19	12.0°C	64 millimetres	204 hours	2.4 days
1920–29	11.2°C	66 millimetres	208 hours	4.3 days
1930–39	11.2°C	64 millimetres	162 hours	2.3 days
1940–49	11.5°C	70 millimetres	201 hours	2.8 days
1950–59	11.4°C	63 millimetres	178 hours	2.4 days
1960–69	11.2°C	73 millimetres	173 hours	2.7 days
1970–79	11.2°C	62 millimetres	184 hours	1.9 days
1980–89	11.1°C	64 millimetres	176 hours	2.8 days
1990–99	11.8°C	48 millimetres	196 hours	1.7 days
2000–04	11.2°C	72 millimetres	186 hours	1.8 days

The ten-year averages show an erratic warming trend throughout the twentieth century, with the last decade exactly 1 degC warmer than the first. There are blips in the trend: the period 1908–22 was remarkably warm, while there was hardly a single warm May between 1972 and 1988, inclusive. Four of the ten warmest Mays of the century occurred between 1989 and 1999 inclusive.

Surprisingly, given the rise in mean monthly temperature since 1900, the trend in sunshine hours has been downwards, the average for the period 1900–09 of 213 hours contrasting with 173 hours in the 1960s and 176 hours in the 1980s. The tendency towards duller Mays appears to have been reversed from 1989 onwards, with several remarkably sunny months in the last 15 years. There is no discernible trend in rainfall figures, although the dryness of the Mays of the 1990s stands out. The 1990s was also the least thundery decade, with fewer than half the number of thunderstorms compared with the 1920s.

May has delivered plenty of meteorological oddments over

the years, with hailstorms and tornadoes, floods and droughts, alongside an occasional foretaste of summer to come and all-too-frequent reminders of winter just past.

Heatwave of the century

In May 1922 the temperature climbed above 20°C (68°F) every day from the 19th onwards, soaring into the high 20s C (over 80°F) between the 21st and 25th. On the 22nd the mercury reached 33°C (91°F) in London, and this remains the earliest date in the season ever to have reached this level. After a brief cooler interlude around the 27th the temperature again climbed towards 30°C (86°F) as the month drew to a close. The rest of the summer was dreadful.

Cold snap of the century

A wicked Arctic outbreak coincided with celebrations to mark the Silver Jubilee of King George V in the middle of May 1935. The 17th dawned clear and frosty – a temperature of –8.6°C (16.5°F) was recorded at Rickmansworth in Hertfordshire – and in many other parts of the UK the day brought several hours' worth of heavy snow, blanketing successively northern England, Wales and finally the West Country. There were snowdrifts 60 centimetres (two feet) deep in the Yorkshire Dales, 10 centimetres (four inches) of snow blanketed the normally mild Wirral peninsula, and level snow lay 12 centimetres (five inches) deep at Tiverton in Devon. Even the usually snow-free Isles of Scilly caught some heavy snow showers during that day.

Drought and flood of the century

These occurred in the same month – May 1989. Parts of the London area measured no rain at all during the entire month, making it the driest May since rainfall records began in the capital way back in 1697. The drought was not severe, though, as the previous two months had delivered a soaking. As so often happens during a hot and sunny month, there were some phenomenal local thunderstorms in various parts of the country. One such hit the Halifax area on the afternoon of 19 May, depositing 193 millimetres (7.6 inches) of rain in two hours at Walshaw Dean reservoir which is located in the Pennines above Hebden Bridge. Put another way, three months' worth of rain deluged the district in just 120 minutes. Flood-waters over a metre (approaching four feet) deep swept through the village of Luddenden, destroying bridges, breaking up road surfaces and damaging homes; the western and southern outskirts of Halifax were badly hit by flooding too.

Freak of the century

One of the best-known British tornadoes of all time carved a path 110 kilometres (70 miles) long through Buckinghamshire, Bedfordshire and Cambridgeshire on 21 May 1950. Worst hit was the small town of Linslade (then in Bucks, now in Beds). Fifty houses lost their roofs, a bakery was destroyed, and farm outbuildings were lifted bodily and dashed to the ground 100 metres away. At Ely a bus was overturned. Mercifully no one was killed.

The first day of spring is . . .

What does the beginning of spring mean to you? To me, it is three things.

It is the yearly battle to persuade journalists that there is no such thing as an 'official' first day of spring. It is trying to remember whether the clocks go forward or back at the beginning of British Summer Time (the American mnemonic 'spring forward, fall back' works best for me). And it is fielding enquiries from all quarters about how unusual it is to get the wintry snap which seems inevitably to accompany the arrival of so-called summer time in the UK.

If you detect a note of cynicism it comes of decades of responding to the same questions. However, the way cold weather returns when the clocks go forward is worth investigating a little further. Of course, in reality there is nothing inevitable about it; it certainly does not happen every year but we do notice inappropriate weather at particular signposts in the calendar. Thus we draw attention to very mild weather at Christmas or rain on the August bank holiday, although climatologically there is nothing the least bit abnormal about either.

The first day of British Summer Time during the last ten years brought drizzle and a gusty wind in 1995, scattered snow showers in 1996, unbroken sunshine in 1997, a very warm but cloudy day in 1998, long sunny spells in 1999, a mixed bag of sunshine and blustery showers in 2000, dull and murky weather with a chill nor'easter in 2001, a dry and bright day in 2002, brilliant sunshine and soaring temperatures in 2003, and cloudy skies but no rain in 2004. In fact we have to go back to 1987 to find a truly wintry day.

If we spread our net a little wider, we do find that bitter winds and snow are not unusual during the last week of March. In 1995 a notable snowstorm hit northern England on the 28th, seriously disrupting transport and bringing down power and telephone lines, especially in West Yorkshire where snow lay 35 centimetres (14 inches) deep at Holmfirth in *Last of the Summer Wine* country, and in the wake of the storm the temperature fell to –10°C (14°F) in northern Scotland.

Bored by the usual questions, I once tried a different tack on the radio. Having consulted my book of ancient weather lore I offered the following during a chat about the weather on Radio 5 Live: 'You know the official definition of spring, don't you? It's when a virgin can step on seven daisies.' Quick as a flash the estimable Jane Garvey shot back: 'But round these parts you just can't find any daisies anymore.' Sometimes it is a pleasure to be trumped.

Spring snowstorm: 4 March 1970

On the front page of the *Daily Telegraph* the summary forecast proclaimed 'scattered sleet or snow showers and sunny intervals'. But even as the paper was being read at countless breakfast tables from Lancashire to London, that forecast was manifestly – embarrassingly – wrong, for snow was coming down in bucketloads over a huge area covering about 60 per cent of England and Wales. The date was Wednesday 4 March 1970.

A forecasting failure of that magnitude would require a government enquiry if it happened today. Thirty-five years ago errors like that did occur, albeit infrequently; nevertheless the forecasters took a lot of flak in the press the following day. What, then, went wrong?

There had been an earlier snowstorm in mid-February followed by a week or so of milder weather, but things turned colder again around the 27th–28th as a chill north wind set in. This was the era when satellite photographs were making their first contributions to weather prediction, and images from the ESSA-8 satellite were available once a day. On 3 March the duty forecaster could see a developing cloud structure south of Iceland which tied in with a small depression centred near Reykjavik on his conventional weather chart. The consensus was that this feature, embedded in the general northerly flow, would track south-south-eastwards across Ireland with little intensification to reach Brittany by the end of Wednesday. In the event the depression travelled slightly to left of the expected track, taking it across the flow which enabled it to deepen abruptly. Thus the deepening depression crossed Northern Ireland on Tuesday night, thence heading across north Wales, the west Midlands, the Thames Valley and the south-east on Wednesday. Snow fell heavily for over 12 hours in a zone 250–300 kilometres (150–200 miles) wide on the cold side – that is the north-eastern side – of the depression track.

This was an exceptional snowfall even by the relatively snowy standards of the 1960s and '70s. A large part of the east Midlands and the northern Home Counties recorded their deepest snow since 1947, nor has it been emulated since. The official weather observer at Thurleigh airfield, some 13 kilometres (eight miles) north of Bedford, was snowed in for 24 hours with hardly any food; he measured level snow 35 centimetres (14 inches) deep by the end of the afternoon, and deep drifts left the local roads completely impassable. The heaviest snow fell in Northamptonshire, where Sywell aerodrome reported a level snow depth of 47 centimetres (18.5 inches). Depths of at least 20 centimetres (eight inches) were found in places as far apart as north-east Wales, north Yorkshire, Birmingham, and east Kent. In the Kent coalfield several hundred miners were stranded with no heat or light for over 12 hours.

In the wake of the snowstorm the northerly winds returned, and temperatures fell to between –10 and –12°C (10–14°F) at night in several districts, and the lowest reading of all was –14.4°C (6°F) at Newport in Shropshire. The snow cover lasted for over a week despite mostly sunny days. They just don't make 'em like that any more . . .

March snow 2001-style

During the first half of March 1970, snow lay on the ground for a total of eight days on the neat lawns of the university campus at Birmingham, which is where I observed that particular snowfall. We were daft enough to build a huge snowman on the roof of the Arts Faculty building, added a scarf, and a fur hat masquerading as a toupee, and named it after our climatology lecturer, Ted Stringer. The remnants of snowman Ted did not finally disappear until the 22nd. Snow lay on the ground for a total of eight days in my Bedfordshire garden during the ten Marches from 1995 to 2004 inclusive, emphasizing how snow-free the month has become in the last decade or so. Nevertheless there was one interesting cold snap in early March 2001, although it did not last long. By the 7th the temperature had climbed widely to 15–16°C (around 60°F).

Before this first sniff of spring, much of the UK was in the grip of a noteworthy spell of wintry weather which lasted well into the first week of March. Heavy snow fell across Scotland and north-east England during the last few days of February, and there were substantial falls across East Anglia on 2 March with 10–15 centimetres (four to six inches) reported from the outskirts of Norwich.

Heavy as it was, the snow was not all *that* unusual, but the extreme cold which followed it certainly was. The ideal recipe for exceptionally low temperatures includes a cloudless sky, an absence of wind, the long winter night, a deep cover of fresh snow, and a geographical hollow or valley. By early March the nights are not quite as long as they are in midwinter, but all the other elements were in place on this occasion. The snow cover is particularly important because snow (especially when it is full of air pockets, as it is when it is newly fallen) is a very poor conductor of heat. That is, when the surface exposed to the air is heated by the sun during the day or cooled by radiation to outer space at night, the change in temperature at the interface is conducted downwards into the snow very inefficiently. Thus on a clear, calm night there is a very rapid drop in temperature at the surface of the snow, and this is immediately transmitted to the layers of air in contact with it. The absence of wind means that this build-up of cold, dense air near the ground is not dispersed by air turbulence, and a valley-bottom site encourages the dense air to collect in a sort of 'pool', rather as water does in a hollow.

Lowest temperatures during the opening days of March 2001 included –16°C (3°F) at Glenlivet, Morayshire, and Biggar, Lanarkshire, on the 1st; –20°C (–4°F) at Carnwath, Lanarkshire, on the 2nd; –22°C (–8°F) at Kinbrace and –21°C (–6°F) at Altnaharra, both in Sutherland, on the 3rd; and –16°C (3°F) again at Altnaharra on the 4th. All these sites are well known 'frost hollows', but it is quite possible lower temperatures occurred elsewhere. At any rate, the reading of –22°C at Kinbrace was within a whisker of the UK's all-time March record of –22.8°C (–13°F) at Logie Coldstone, Aberdeenshire, on the 14th in 1958. Many climate experts were surprised to see such low temperatures in this rather warmer era, but the fact is that there were many fewer weather-recording stations in the highlands before about 1990, and it is quite possible that readings appreciably lower than –22.8°C might have been logged had instruments been present at Kinbrace and Altnaharra in 1958.

Walking in the hills

Scarcely a single spring passes without newspaper reports of people walking in Britain's hill-country being caught out by a sudden change in the weather. All too often someone loses his or her life. Even when there are no fatalities mountain rescue teams and airborne search-and-rescue units are called out at great expense, and often at considerable risk to their own lives.

It is a sad fact that the majority of these dramatic and sometimes disastrous events could be avoided if only the people involved listened to the experts before heading out into the hills. Even better, a little self-education on the meteorology of our upland regions would give first-time hill-walkers some knowledge of what to expect. It is no coincidence that the majority of rescues are of 'townies' who venture into the hills on a whim.

It is essential to take advice from those who know the district. First of all, get hold of an appropriate weather forecast. I don't mean listen to the girl on the telly, or look at the maps in the newspaper; these days there are specially written detailed forecasts available on premium-rate telephone and fax lines, and also on the Internet and mobile phone networks, designed for those heading for the hills. Some districts also have a local forecasting office where it is still possible to speak to a human meteorologist. Do not take the forecast as absolute gospel; rather, treat it as a reasonably accurate guide. Similarly, check for any avalanche warnings which may be in force, or predicted, for the district. Apart from seeking weather information, experienced walkers will also tell you to ensure you are suitably clothed and shod, have appropriate food and water, and carry a compass and mobile phone. It is also essential to leave an itinerary or map of your projected walk at the local mountain rescue centre or, if there is none, with a responsible friend or relative. All this may sound like common sense, but it is extraordinary how many folk are rescued from the deep snow and severe wind-chill of a mountainside clad in just a Marks and Sparks anorak and the latest fashion trainers.

And what of the weather? Spring is probably the most dangerous season of all for the casual walker in, for example, the Scottish highlands, the Lake District, the Pennines or Snowdonia. The first warm day of the season will draw forth thousands of city-dwellers, blinking in the spring sunshine rather like moles waking from their winter hibernation. They head straight for moor and mountain expecting the sun to be shining just as warmly; instead they find winter still firmly in charge.

Meteorologically speaking, altitude changes everything. Climb a mountain and this is what happens: down go air pressure, temperature, sunshine hours, visibility; up go wind speed, cloudiness, rainfall and snowfall, and humidity. And, of course, the combination of lower temperature and a stronger wind means that the wind-chill factor increases sharply. Changes in the weather are also more sudden and more dramatic than they are at lower levels. With all this happening it is easy to see how an inexperienced hill-walker can get caught out.

On most days, the temperature falls systematically with height above sea level. On average the rate of decline in the British Isles is 1 degC every 175 metres (1 degF every 300 feet), but it can vary quite widely either side of this average. For example, in an unstable polar airflow, brought to us on a blustery north-westerly wind with a mix of sunshine and showers, the rate of decline is more like 1 degC per 105 metres (1 degF per 180 feet).

Imagine, for a moment, waking up on such a morning in Aberdeen, round about Easter; the sky is blue, the atmosphere is clean and pure, the air is mild although there is a fresh breeze, and the temperature stands at a near-average 10°C (50°F) – the perfect day for some healthy exercise. You jump in the car and head inland, aiming perhaps for Braemar or Aviemore. At Aviemore, some 230 metres (750 feet) above sea level, the sky is part-cloudy and the temperature is 8°C (46°F), but the snow-clad slopes of the Cairngorms gleam brightly in the fitful sunshine. Turning left onto the B970 you cross the River Spey, decide against the delights of the Whisky Centre at Inverdruie, and head through Glenmore Forest, eventually negotiating the hairpins as you climb towards the Cairngorm car park. There is a distinct nip in the air, you notice, as you lock the car, adjust your anorak, and re-tie the laces of those expensive trainers. You are now at 600 metres (2,000 feet) above sea level, the temperature is 4°C (39°F) and the wind is keen. No matter, a brisk walk will warm you up.

The main footpath paralleling the chair-lifts on the ski slopes is easiest, and in an hour or so you are crunching through snow as the Ptarmigan restaurant comes into sight. Already you have endured one or two brief but very sharp showers of hail and sleet, myriad tiny frozen missiles driven horizontally by the near-gale that is now blowing. At the restaurant, sadly closed today, you are just over 1,000 metres (roughly 3,500 feet) above sea level, the temperature is –1°C (30°F). The wind cuts through your woefully inadequate clothing like a knife, but your map shows you that the summit of Cairn Gorm is less than a kilometre (half a mile) away, before that map is snatched from your hands by the gale and disappears in the general direction of Arbroath. Standing on the top of Cairn Gorm always was one of your ambitions, and it is, after all, only four o'clock . . .

You have a choice. Go back now and you will probably make it to the car park suffering no more than mild hypothermia and exhaustion. Carry on and you will likely end up just another statistic.

There is an electronic weather station on the mountain summit which is 1,245 metres (4,084 feet) above the sea. At four o'clock that afternoon it registered a temperature of –4°C (25°F), and a sustained wind speed of 70 knots (80 mph) gusting to 90 knots (103 mph).

Clout-casting in May

Casting clouts is, according to the old saying, inadvisable before the end of May. Sometimes it seems as though much of that old country lore was designed by a bunch of busybodies to discipline our forebears' everyday lives, and if they had had their way no doubt such impetuous activity would have been totally prohibited: 'For venturing forth on 29 May without an overcoat you are fined tuppence, to be paid at the next sitting of the court leet.'

But does that ancient proverb

Cast not a clout ere May be out

actually mean what it says? Some experts believe that the 'may' in the proverb should be spelt with a small 'm' because it refers not to the month, but to may blossom – the flower of the hawthorn. As supporting circumstantial evidence they point to the fact that haws and hawthorn along with blackthorn, whitethorn and other common wild-growing bushes, appear time and time again in the lexicon of country weather lore.

This option could make more sense than the more obvious one: may blossom appears on average in late April in the warmest parts of southern England but not until the end of May towards the coast of Northumberland, Durham and North Yorkshire, and even later than that over the hills of northern England and in many parts of Scotland. It is illogical that clout-casting and similar activities should begin on 1 June over the entire country, given the widely varying climate of our small island. Against that, it could be argued that it is equally illogical to switch to summer clothing in the middle of April simply as a consequence of an early heatwave, when we all know that dramatic warm spells in spring are often followed by equally dramatic cold spells.

The truth probably lies somewhere between the two. It is quite likely that different people in the same village used the proverb differently, and, given the fickleness of human nature, it is not difficult to imagine the same wizened old sage asserting one meaning in one year, and the other in another.

What can a meteorologist make of all this? Is there anything relevant that we can absorb from this particular old saying? Many meteorologists look at weather lore with too literal an eye, see nothing which relates to their science, and laugh patronizingly at our ill-educated ancestors. They would, I think, do better to take a couple of steps back, and to ask what the point was of all these old sayings.

They date back to times when Britain was a largely rural country, and most people who worked on the land could neither read nor write. The only real way of passing knowledge from generation to generation was by word of mouth, and such information was most readily assimilated when linked to events in the natural calendar, such as the equinoxes and the months, and also to dates in the church calendar – annual festivals like Lady Day, Candlemas, Lammas, All Hallows and Christmas, and also saints' feast days such as those of Swithun, Stephen, Martin and Luke. Proverbs and rhymes are easy to remember and to pass on, and simple rhyming couplets such as the one quoted above are probably most persistent of all; it is some of these which have lasted so well into the twenty-first century.

Rather than take these sayings literally, then, we should just absorb the broad underlying message. What the rhyme about casting clouts in May really means is this. Spring in Britain does not advance in a straightforward way, each week a little warmer than its predecessor; rather, it comes in a series of fits and starts, and a typical spring will bring short warm spells interspersed with sudden reversals in the weather. By the time May arrives it is easy to be lulled into a false sense of security by a period of warm sunny weather, a sense that summer has now arrived and that the cold weather has been banished once and for all. The rhyme merely reminds us that this is not true, that damaging frosts are still possible for some weeks to come. There is a secondary feature of late-spring weather which may be part of the original message: even during the warmest May weather, the early mornings and the evenings are usually decidedly chilly – those balmy evenings of high summer are actually quite rare this early in the season.

A meteorologist would also point out that, even in an average year (if such a thing exists), sudden cold snaps can follow brief heatwaves until at least the middle of June. It is not really until after 20 June that the frequency of cold days diminishes sharply. The same is true of those warm sunny days with chilly mornings and evenings. Part of this discrepancy is probably due to the 11-day slippage when we changed from the Julian to the Gregorian calendar in 1752. Again, though, we shouldn't take too literally the implication of the old saying that things automatically change as May gives way to June.

One recent example illustrates just how fickle the weather can be in Britain at this season. May 1995 began with a spell of record-breaking heat and sunshine, the temperature climbing to around 27°C (80°F) on every day during the first week. Within five days northerly winds brought heavy snow showers to Scotland and northern England with widespread night frosts even in southernmost counties of the UK, and the temperature slumped dramatically to –6°C (22°F) at Saughall in Ayrshire overnight 10th/11th. On the 17th the maximum afternoon temperature was 7°C (45°F) in parts of southern England and just 4°C (39°F) at Fylingdales in Yorkshire. Another brief heatwave followed between the 22nd and 24th, with 25°C (77°F) reached locally, but a sluggish northerly airflow set the scene for further early-morning frosts on 1 and 2 June.

Spring frosts and frost hollows

Keen gardeners and horticulturalists are always concerned about spring frosts, and many spells of fine and settled weather in March or April, and occasionally in May as well, will bring overnight frosts as well as warm sunshine by day. Not everywhere suffers frost to the same degree, however, and it is during quiet periods like this that the contrasts between frosty valleys and frost-free hilltops are most evident. Anyone who lives in a frost hollow knows exactly how rapidly the temperature can drop after dark, and how frequently frozen car windscreens have to be scraped clear even after the spring equinox.

When the sky is clear and the wind is light, the ground radiates heat energy to outer space both by night and by day. During the day this loss is more than made up by the incoming energy from the sun, but at night it means that the ground quickly gets cold, and the air in contact with the ground also cools down. Cool air is denser than warm air so it tends to drain slowly down slopes and to collect in valleys and hollows, as long as there is no significant wind to mix up these lower layers of the atmosphere.

Probably the most famous British frost hollow of all is in Rickmansworth, Hertfordshire, where a railway embankment effectively dams the valley and allows a pond of cold air to develop very readily. This particular spot gained its notoriety thanks to an investigation in the 1930s by the meteorologist Eric Hawke, who demonstrated that under ideal conditions the temperature in that Rickmansworth valley can be as much as 15 degC (27 degF) lower than in central London.

There are scores of similar frost hollows in the chalk downlands of the Home Counties where the flow of cold air down the length of the valley may be impeded by woodland, built-up areas, road and rail embankments, and so on. Readers who live in the lower-lying parts of Berkhamsted, Chesham, High Wycombe and Dorking will be intimately familiar with this phenomenon. Studies have also examined similar frost-prone localities elsewhere in the UK; the most prominent of these include Houghall near Durham, Newton Reigny near Penrith, Alston in Cumbria, Rhayader in mid-Wales, Newport in Shropshire, Braemar in Aberdeenshire, and Altnaharra in Sutherland, but there are undoubtedly hundreds of others which have not been examined.

Not even coastal districts, often frost-free thanks to the warming influence of the adjacent sea, are immune from the frost-hollow effect. Notably cold such places include Gogerddan near Aberystywyth, and Litton Cheney in west Dorset.

The Tibshelf tornado

Tibshelf, an unassuming Derbyshire mining village 26 kilometres (16 miles) north of Derby and ten kilometres (six miles) west of Mansfield, hit the headlines just over 50 years ago. On 19 May 1952 a small but destructive tornado hit the district around four o'clock in the afternoon. The tornado cut a swath three kilometres (two miles) long through the village, causing considerable damage to property, trees and crops; fortunately no serious injuries were reported.

May 1952 was very warm, and a mid-month heatwave culminated on the 19th in a temperature of 30°C (86°F) in London. A humid south-easterly airflow from the continent covered the Midlands, but during the afternoon the warm air was squeezed from two separate directions. A cool east wind from the North Sea blew across Lincolnshire (21°C or 70°F at Lincoln itself), and a colder north-westerly flow from the Irish Sea (17°C or 63°F in Manchester) travelled across the Pennines and the Peak District. The instability caused by the convergence of these three contrasting air-masses provided the impetus for the growth of the thunderstorm which triggered the tornado.

An excellent eyewitness account was provided by Mr W. H. Hill of Lane End, just north of Tibshelf:

> The calmness was suddenly broken by a single lightning flash and crack of thunder. The dog which was with us dashed home at once, and we immediately left the field on a tractor. On reaching the farm a great roaring noise, something like the continuous roar of a great waterfall, could now be heard. To the west . . . we could

> see all the trees in violent turmoil, branches wrenched off, flung upwards, and whirled around with other debris. As sheds and loose objects from the gardens began to sail over, we took shelter in the house.
>
> The air was quite still until the tornado struck the house, flinging a piece of corrugated iron through the window. I was able to get a full view of the vortex, which roared like the updraught in a gigantic flue. The top of a large wooden building a few yards from me suddenly burst upwards and every piece of loose wood was torn out. A large cornstack nearby was whipped into the air sheaf by sheaf as by a giant juggler, and held there before being flung aside. Heavy stone ridge tiles were plucked from the house, and the whole yard was full of flying stones and rubble as the disturbance passed over, leaving the air as quiet as before.

To survive such a close encounter with a tornado must be quite rare, but to survive it and be able to put together such a clear and measured description of the event is probably unique.

The tornado travelled from north to south with the general airflow – most British tornadoes move from south-south-west to north-north-east – and seriously damaged more than 100 homes in and around Tibshelf. Many outbuildings were completely demolished, dozens of trees were uprooted, but mercifully only one person was injured.

Chapter 4

The four seasons: summer

Introduction

Summer for the astronomers begins around 22 June and ends on 22 September give or take a day, according to our position in the leap year cycle – in other words, from the summer solstice to the autumn equinox. Many have pointed out the illogicality of this scheme, given that it requires the last day of summer to have approximately the same hours of daylight as the last day of winter, as both fall at the equinoxes. They suggest that it would make more sense for the seasons to be *centred* on the solstices and equinoxes rather than separated by them.

Fortunately, we have other ways of dividing the year. The old ecclesiastical calendar told us that summer began on 25 May (St Urban's Day) and ended on 23 August (the day before St Bartholomew's Day) – *Aestuat Urbanus, autumnat Bartholomaeus* – but in popular secular parlance, May, June and July were regarded as the summer months in the Middle Ages. Now that most of us live in towns and cities, and the majority of people prefer August as the month for summer holidays, the months of June, July and August are almost universally regarded as summer. This coincides with the climatological quarter used for statistical summaries and analyses.

Professor Lamb divided the year – or, rather, he considered that the year divided itself – into five *natural seasons*. Two of these seasons straddle the popular perception of summer. We have already seen that *late spring and early summer* runs from 1 April to 17 June on average, and Lamb's *high summer* encompasses 18 June to 9 September.

June in the twentieth century

'June is bustin' out all over, all over the meadow and the hill', wrote Oscar Hammerstein in one of the best-known numbers in the classic musical *Carousel*. The action was set in northern New England where the early summer climate is sufficiently similar to that of the old country for us to question Hammerstein's grasp of elementary climatology. The average June on the Maine coastline is just as likely to be a fickle month with occasional downpours and a cold wind off the sea as it is at the English seaside.

The statistics certainly support the notion that changeable, rainy weather often sets in during the middle of June following a dry spell. When climatologists came to analyse the reason why, they discovered that unsettled westerly winds, rare between mid-March and mid-June, showed a marked increase in frequency during the second half of the month. This is why Wimbledon, Ascot and all the other high-profile sporting events scheduled in this part of the summer are so often spoilt by the weather, and anyone who expects June to live up to its 'flaming' epithet will normally be sorely disappointed. Looking back through the twentieth century, only ten Junes could reasonably be described as predominantly warm and sunny, which leaves ninety which weren't.

The sun may be above the horizon for 16 hours per day in southern England and more than 18 hours in northern Scotland, but it cannot make its presence felt if the skies are full of cloud. Furthermore, the seas around our shores are, at the beginning of June, still three months away from their warmest, and the wind has to blow across the sea to reach us. Even the relatively narrow English Channel can have an effect: a warm and humid southerly airflow can be sufficiently cooled and moistened in its brief passage across the water to leave your Brightons and Bournemouths bathed in mist and drizzle. The North Sea and Norwegian Sea remain particularly cold in June, and an insistent northerly or north-easterly wind may hold the temperature close to 10°C or even lower right through the day.

The decade-by-decade statistics for June illustrate how the month's climate has varied since late Victorian times. These figures are an approximate geographical average over the whole of England and Wales:

Decade	*Mean temperature*	*Rainfall*	*Sunshine*	*Thunderstorms*
1880–89	14.0°C	59 millimetres	199 hours	–
1890–99	14.6°C	61 millimetres	195 hours	2.3 days
1900–09	13.7°C	69 millimetres	185 hours	3.4 days
1910–19	14.1°C	61 millimetres	203 hours	2.6 days
1920–29	13.7°C	53 millimetres	205 hours	2.1 days
1930–39	14.7°C	61 millimetres	201 hours	3.8 days
1940–49	14.6°C	55 millimetres	204 hours	3.1 days
1950–59	14.4°C	66 millimetres	186 hours	3.3 days
1960–69	14.6°C	60 millimetres	204 hours	2.8 days
1970–79	14.1°C	58 millimetres	189 hours	3.0 days
1980–89	14.1°C	73 millimetres	177 hours	3.6 days
1990–99	14.2°C	73 millimetres	180 hours	3.0 days
2000–04	15.1°C	56 millimetres	200 hours	1.4 days

Unlike the first five months of the year, the June statistics show neither an upward trend in mean monthly temperature during the twentieth century, nor a dramatic warming during the 1990s. The warmest decades were the 1930s, the 1940s and the 1960s, whereas the 1900s and 1920s were markedly cooler than any other decade. There is, though, a strong hint in the last line of data that the warming trend seems finally to have established itself during the first years of the new century; there is also a suggestion that June has started to become drier, sunnier and less thundery, in line with the more established trend in July and August. The range between the coolest and warmest decades was smaller in June than in any other month.

Nor was there any obvious pattern in the rainfall and sunshine records for most of the century, although the cloudy and wet Junes during the 1980s and 1990s stand out. There were more very wet Junes between 1980 and 1999 than there were in the preceding 100 years. There is one other curiosity worth noting here: the cool Junes of the 1920s were also relatively dry and quite sunny too, contrary to the general summer relationship that cool months are usually dull and wet while warm months are normally sunny and dry.

June has provided us with a host of extremes over the years, and a handful of strange-but-true events as well. There have been hailstorms and tornadoes, floods and droughts, heatwaves and cold snaps, and even one day when it snowed.

Heatwave of the century

There is only one contender for this title: June 1976. This heatwave was the climax of the hottest summer for almost three centuries, and it lasted from 22 June until the middle of July with the temperature widely reaching 27°C (into the 80s F) or more on each of 25 consecutive days. Even more remarkable, the mercury soared to 30°C (86°F) on 17 days, and to 32°C (90°F) on 15 successive days between 23 June and 7 July. No heatwave before or since has produced more than five consecutive days above 32°C. In many places the hottest days of the entire summer were 27 and 28 June, and on the latter date a maximum temperature of 35.6°C (96.1°F) was recorded at Mayflower Park in Southampton.

Cold snap of the century

Queen Elizabeth II's coronation, on 2 June 1953, was meticulously scheduled on the basis of advice from the Meteorological Office. According to government meteorologists at the time, 2 June was one of the most reliably dry and sunny days during the entire year, statistically speaking. Now, statistics may be able to describe the British climate over a long period, but they are a fat lot of use when it comes to forecasting the weather. Coronation Day turned out to be one of the coldest and most miserable June days of the entire twentieth century. Leaden skies produced fitful rain and a chill north wind, and the midday temperature in London was just 9°C (48°F). At least the capital escaped the torrential downpours which hit north-east England that day, with upwards of 50 millimetres (2 inches) of rain falling locally in Northumberland.

Flood of the century

Rain fell without a break for 59 hours in London during June 1903, from one o'clock in the morning on the 13th until midday on the 15th. In fact it rained heavily on each day from 8 to 20 June inclusive, and during this 13-day period 184 millimetres (7.25 inches) of rain fell in the capital, and 220 millimetres (almost nine inches) in parts of north Surrey and north-west Kent, with Carshalton, then a quiet village at the foot of the North Downs, topping the rainfall league table. Streets were awash, and large sections of the London Underground were shut down when water flooded the tunnels.

Freak of the century

It has to be Monday 2 June 1975, when snow fell over much of Britain; in fact, snow and hail showers were observed as far south as Devon and Hampshire. Some three centimetres (over an inch) of snow lay on Buxton cricket ground that morning, resulting in the abandonment of the county championship match between Derbyshire and Lancashire, while in the Aberdeenshire hills there were reports of upwards of 15 centimetres (six inches) of level snow. Never before nor since has snow fallen as widely in June, yet by the end of the week blistering sunshine sent the temperature rocketing into the high 20s C (over 80°F).

July in the twentieth century

According to the old saying, the British summer comprises 'two fine days and a thunderstorm'. Time and again over the years, the chief forecaster's July weather charts show a big fat high-pressure system developing over the British Isles promising some real summer weather. By day two the 'high' has drifted over the Low Countries, allowing a gentle southerly flow of Mediterranean origin to send the temperature soaring into the high 20s C (the 80s F). Then on day three the highest pressure has migrated to Germany or the Baltic, very moist air from the Bay of Biscay insinuates itself across southern Britain, temperatures head for the 30°C (86°F) mark, and the combination of heat and humidity triggers widespread thundery activity. Thereafter cooler westerly winds return and the abbreviated heatwave is over.

Sometimes, though, July does deliver. Day after day of blue skies, blazing sunshine and soaring temperatures: such Julys are rare enough to enter the communal memory. People of different generations will recall (or used to recall) the summers of 1911, 1921, 1933, 1959, 1976 and 1995, all of which featured a hot and sunny July. Occasionally, of course, we get the other side of the coin, with cloud-laden skies, repeated downpours, relentless wind, and subdued temperatures. Funnily enough we remember these summers without being able to put a name – or number, rather – to them. But the record books show them up: 1912, 1920, 1946, 1954, 1965 and 1988.

There is probably a feeling abroad that good Julys have become more frequent in recent decades, while truly bad Julys are now very rare. A statistical analysis of the Julys of the last century and a quarter will help us determine whether there has been any underlying trend in Britain's July climate:

Decade	*Mean temperature*	*Rainfall*	*Sunshine*	*Thunderstorms*
1880–89	15.5°C	95 millimetres	193 hours	–
1890–99	16.0°C	73 millimetres	192 hours	3.3 days
1900–09	15.6°C	68 millimetres	204 hours	3.0 days
1910–19	15.5°C	73 millimetres	177 hours	3.4 days
1920–29	16.1°C	86 millimetres	193 hours	2.8 days
1930–39	16.3°C	86 millimetres	178 hours	3.9 days
1940–49	16.5°C	70 millimetres	190 hours	3.7 days
1950–59	16.1°C	78 millimetres	182 hours	3.9 days
1960–69	15.4°C	77 millimetres	174 hours	3.8 days
1970–79	16.1°C	67 millimetres	181 hours	2.3 days
1980–89	16.5°C	61 millimetres	188 hours	2.7 days
1990–99	16.9°C	55 millimetres	206 hours	2.0 days
2000–04	16.4°C	75 millimetres	173 hours	1.8 days

The statistics show that July, like most other months of the year, has become markedly warmer during the last 25 or 30 years. The twentieth century opened with two decades of cool Julys followed by a warmer episode which peaked in the 1940s with a mean monthly Central England Temperature of 16.5°C (61.7°F). There was then a marked deterioration which culminated in the 1960s having a mean temperature of 15.4°C (59.7°F), making this the coldest decade since the 1690s. The warming since then has been dramatic and the 1990s averaged 16.9°C (62.4°F), warmer than any other decade in the 340-year-long record. Since 2000 the upward trend in July temperature has been checked, though this may be only a temporary setback.

July has also become a much drier, sunnier and less thundery month than it was several decades ago. The rainfall averaged over England and Wales during the 1990s was just 55 millimetres (2.16 inches) compared with 86 millimetres (3.39 inches) during the 1920s and '30s, while sunshine amounts have improved by almost 20 per cent in the last 40 years; in the 1990s July actually overtook June to become the sunniest month of the entire year. As with temperature, though, the last few years have seen a reversal, possibly temporary, of the trend towards drought and sunshine.

July has provided us with some remarkable record-breaking weather over the years, including some bizarre occurrences as well as the usual mix of heatwaves and cold spells, floods and droughts, thunder, hail and tornadoes.

Heatwave of the century

July shares the twentieth century's biggest and longest heatwave with June – the 1976 one. There was plenty to go round: this was arguably the most remarkable spell of high temperatures in the UK for 300 years, and perhaps for much longer even than that. The two hottest July days of the twentieth century were the 2nd and 3rd in 1976, when the temperature touched 35.9°C (96.6°F) at Cheltenham in Gloucestershire. These days occurred in the middle of the prolonged heatwave which began on 22 June and lasted until 16 July. 2 July also saw the temperature climb to 32.4°C (90.3°F) at Wauchope Forest in Roxburghshire, equalling the Scottish record up to that time.

Cold snap of the century

In the midst of the warm 1990s, Britain experienced one July day of truly exceptional coldness. On 9 July 1993 a cold front moved southwards across the country, introducing air which had travelled from far inside the Arctic Circle. During steady rain in the rear of the cold front the temperature dropped sharply, and at lunchtime the mercury at Whipsnade in Bedfordshire stood at just 7.5°C (45.5°F), some 13 degC (23 degF) below the seasonal average. It will be difficult to locate this outstanding day in future record books because the day's maximum temperature, an unremarkable 16 or 17°C (61–63°F) in central and southern Britain, occurred around breakfast time or just after. Further north, in Scotland and northern England, the sun came out in the afternoon, lifting the temperature there to 14–16°C (57–61°F).

Downpour of the century

July 1955 provided a splendid example of the contrariness of the British climate. Taking the country as a whole it was one of the driest and sunniest months of the entire century, and most parts of the UK were very warm as well. On the one hand a large swath of England recorded negligible rainfall, and on the other a new all-time UK record for rain in one day was established on the 18th. On that date 279.4 millimetres (exactly 11 inches) of rain fell at Martinstown, near Dorchester, much of it within a ten-hour period. Put another way, four months' worth of rain fell in less than half a day, and that is an awful lot of water in a very short time. Although a substantial proportion of the rain was absorbed by the underlying chalk, serious flooding followed in several parts of west Dorset, including the towns of Dorchester and Weymouth.

Legend of the century

St Swithin, whose feast day falls on 15 July, has arguably given us our best-known piece of ancient weather lore. Swithin, or more properly Swithun, was a ninth-century Bishop of Winchester who left instructions to be buried in the churchyard 'where the sweet rain of heaven may fall on my grave', rather than in the customary shrine inside the cathedral. According to the ancient lore, if it rains on Swithin's Day it will rain for 40 days thereafter. It will come as no surprise that there was not a single example during the twentieth century when the old saying came true. Some climatologists, however, argue that we should take a broader view: research during the middle of the last century showed that weather patterns that become established during the height of summer, around mid-July, say, tend to persist for several weeks. Perhaps this is what the anonymous authors of the Swithin legend had recognized.

August in the twentieth century

> Never have I asked an August sky,
> 'Where has last July gone?'

Hammerstein, again, this time in *Oklahoma*. The character, Laurey, refuses to fret over a lost love, reasoning that there are plenty more pebbles on the beach. August, though, sometimes brings summer to a premature end, and whether you are in Oklahoma or Oxford the days get perceptibly shorter and the evenings darker as the month progresses.

Long ago when Britain was an essentially rural country and the vast majority of the people worked on the land, August was regarded as the first month of autumn. Dictionaries are unclear about the origin of the word 'autumn', but it is interesting that the German equivalent, *Herbst*, shares a root with our word 'harvest'. There is no doubt that in medieval England August was an autumn month *because* it was the first of the harvest months. Try telling your typical present-day urban Brit that autumn begins on 1 August and he will probably laugh in your face. Or perhaps not, since he is more likely to be roasting on some Mediterranean beach.

Few would now demur from the generally held view that

August is certainly part of summer in the British Isles, although towards the end of the month there are occasional signs of approaching autumn, especially in Scotland where the nights draw in more rapidly than they do further south. These longer nights may produce mists, fogs and heavy dews when the weather is settled. When it is not, a blustery wind blowing from the west or north-west may bring a distinctly autumnal chill to the air, and the first falling leaves can be seen swirling along the pavements of northern cities like Aberdeen or Inverness. On the coldest late-August days of all, the afternoon temperature may fail to reach 13°C (55°F) even in southern England, and reach only 10°C (50°F) in the north of Scotland, while the Cairngorm summits may briefly sport a glistening dusting of snow. Such days, though, are very much the exception.

Averaged over a long period, the decline in temperature between the first and last days of August is almost imperceptible, and heatwaves can occur at any time. Maxima of 32°C (90°F) or more have been recorded somewhere or other in the UK on every single date during the month. Britain's highest temperature of the twentieth century occurred on 3 August 1990 when 37.1°C (98.8°F) was observed at Cheltenham, and this record was smashed out of sight on another August day, the 10th in 2003, when 38.1°C (100.6°F) was reported at the Royal Botanic Gardens at Kew, south-west London. Even as late in the month as 31 August 1906, the mercury soared to 35°C (95°F) at Maidenhead in Berkshire.

Decade	*Mean temperature*	*Rainfall*	*Sunshine*	*Thunderstorms*
1880–89	14.9°C	71 millimetres	175 hours	–
1890–99	15.5°C	89 millimetres	187 hours	2.9 days
1900–09	15.0°C	90 millimetres	194 hours	2.8 days
1910–19	15.6°C	94 millimetres	166 hours	3.1 days
1920–29	15.0°C	90 millimetres	178 hours	3.0 days
1930–39	16.3°C	68 millimetres	179 hours	3.4 days
1940–49	16.1°C	79 millimetres	179 hours	1.6 days
1950–59	15.7°C	99 millimetres	169 hours	3.5 days
1960–69	15.2°C	88 millimetres	157 hours	2.3 days
1970–79	16.0°C	73 millimetres	171 hours	3.0 days
1980–89	15.8°C	76 millimetres	181 hours	2.8 days
1990–99	16.8°C	68 millimetres	201 hours	1.8 days
2000–04	17.3°C	81 millimetres	191 hours	2.4 days

The changes in August temperature during the twentieth century mirror the pattern observed in many other months during the year, especially July. The first three decades were comparatively cool, the 1930s and 1940s were much warmer, but it turned cooler again between 1950 and 1988, with a complete absence of warm months between 1960 and 1972. The century ended with a decade of exceptional warmth, not matched at any other stage of the 340-year-long Central England Temperature record, let alone during the twentieth century, and the warming trend (unlike that in July) has shown no sign of easing during the early years of the twenty-first century. There seems to be a correlation between temperature and rainfall during August, with the warm periods tending to be dry whereas the cooler ones were mostly wet; the warmest five decades averaged 73 millimetres (2.88 inches) of rain, taking the country as a whole, while the coolest five decades averaged 92 millimetres (3.63 inches). Superficially, this association may seem obvious, but from a meteorological point of view it is not logical. Warmer air can support more moisture than cold air can, and our heaviest summer downpours frequently come during thundery weather when both temperature and humidity are high. Thus one might have expected that warmer Augusts would also be wetter, on average, and this was indeed the case in August 2004, a month which was both very warm and so wet that it heavily skews the rainfall figure for 2000–04. Sunshine totals showed little correlation with either temperature or rainfall. However, the cool 1960s were also very cloudy, while the highly abnormal 1990s were not only very warm and dry but also much sunnier than any other decade.

August has, as one might expect, brought some of our most exceptional heatwaves and droughts, but it has also provided us with some outstanding rainstorms, floods and thunderstorms.

Heatwave of the century

Most people wilted. It was exactly ten years before the century ended, and Britain was enduring – a few hot-weather addicts insisted they were enjoying – the most intense heatwave Britain had had since reliable temperature records began. The temperature climbed to 32°C (90°F) or more on the first four days of August 1990, climaxing on the 3rd with a new all-comers' UK record of 37.1°C (98.8°F) at Cheltenham in Gloucestershire. This remarkable spell of heat and humidity has already been eclipsed, just three years into the new century, by the 'killer heatwave' of August 2003.

Flood of the century

This was a most difficult choice. Torrential downpours had brought serious flooding to Norfolk in August 1912, to Somerset in August 1924, to north Devon in August 1952 and to London in August 1975. But when it comes down to it, the Lynmouth flood of 15–16 August 1952 was clearly the worst, because 34 people lost their lives. Over 250 millimetres (ten inches) of rain – three months' worth – fell in 22 hours over Exmoor, and torrents of water cascaded down the East and West Lyn Rivers on the precipitous northern flank of the moor. The two rivers converge in the village itself which suffered extensive physical damage as a result. The Boscastle flood of August 2004 brought back memories of Lynmouth, but in truth the more recent event was less destructive and no lives were lost.

Gale of the century

High winds are very rare during the summer, and gales are hardly ever reported in August. But windy weather does sometimes catch out weekend sailors who think that August is a safe month, and they do not bother to check the Shipping Forecast. The gale that struck our southern and western waters on 13–14 August 1979 was an abnormally violent one, and it coincided with that year's Fastnet Race – the culmination of the Admiral's Cup. Sustained winds of 50 knots (approaching 60 mph), gusting to 70 knots (80 mph) or more, in the South-west Approaches were inadequately forecast, and the race was thrown into chaos. Fifteen participants were drowned, 136 were rescued, and of the 303 yachts which left the Solent three days earlier only 85 finished.

How hot is hot?

How hot is hot? Have you ever listened to a weather forecast on the wireless and found yourself disagreeing strongly with the forecaster? Maybe a prediction of, say, 'rather warm' conflicted with your impression that it was damned hot and excessively humid to boot.

Temperature is not as simple a matter as we think. Today we are bombarded with temperatures: in radio and television forecasts, the newspapers, digital displays attached to office blocks or in banks. All these keep us up to date with current temperature readings or those predicted for the next three or four days. It seems that we all need to know how hot or cold it is going to be. Oddly, though, human beings are remarkably poor thermometers. In a recent scientific exercise, a group of people were taken outside and asked to guess the shade temperature. Only 42 per cent of the guesses were within 3 degC (5 degF) of the actual temperature, 64 per cent were within 5 degC (9 degF), and 5 per cent were more than 10 degC (18 degF) adrift.

One reason for these levels of inaccuracy is that we actually feel heat and cold by means of our physiological responses to our surroundings, in particular to the speed with which we gain or lose heat energy from the air around us. Physiological responses vary from individual to individual, so one person's 'lovely warm day' may be another's 'hot, sticky and uncomfortable' one. And they also vary from day to day in any one individual, according to health, tiredness, level of activity, and type and thickness of clothing.

Several external factors in addition to the air temperature also influence how hot we feel. These include the humidity level (which, confusingly, itself depends to a certain extent on the temperature, because the warmer the air is the more moisture it can hold), the wind speed and gustiness level, the strength and directness of sunshine, and pollution concentrations. Thus our bodies tell us that it is 'hotter' when a high temperature is accompanied by high humidity, strong sunshine and no wind. Conversely we feel colder when a low temperature is accompanied by high humidity, no sunshine and a strong gusty wind. A high pollution level affects our health and makes us less able to cope with extremes at both ends of the spectrum.

Variations in humidity are particularly important. Even in moderate heat with the air temperature between 21 and 26°C (in the 70s F), they have an impact on our lives. When the air is both hot and humid the weather feels sultry and oppressive. Our reactions slow down, we feel lazy, and our efficiency at work suffers. This is particularly true of hard, physical, outdoor work, but it applies also to people who work in factories, offices and shops without the benefit of air conditioning. We should remember, too, that high humidity levels bring sticky, airless nights which in turn cause most of us to have poor or disturbed sleep, leading to tiredness and inefficiency at work.

By contrast, moderately dry air is stimulating, leading to high work efficiency. The temperature usually drops briskly after dark, we get a better night's sleep and we are more likely to feel fully refreshed the following morning. But when the humidity drops to a very low level and the air becomes excessively dry, the effects are again negative. People become dehydrated without realizing it, and this leads to irritability and lack of concentration.

At low temperatures the air holds comparatively little moisture, so the effect on the body of increasing or decreasing humidity is much smaller than it is in hot weather. Even so, we can detect some differences. We describe cold, damp air as 'raw', and cold, dry air as 'bracing'. A raw wind is the more unpleasant of the two because it makes our clothing and any exposed parts of the body damp, and in response to this we expend more energy trying to counteract these effects. In other words it is harder to keep warm. A bracing wind, especially if the sun is shining, can be countered by a brisk walk. However, excessively dry air at low temperatures leads to dry, cracked skin, chapped lips and sore eyes, (and also, incidentally, to an accumulation of static electrical charge leading to those nasty little shocks whenever one touches a metal object).

The lowest temperatures we are likely to encounter occur

when there is no wind at all. In these cold but still conditions, suitably clothed, healthy human beings normally cope very well. In effect, we develop our own microclimate, warmed by body heat, within our clothing. Fur and feathers provide the same sort of insulation for mammals and birds – and for some people too.

Heat, humidity and sunstroke

Coping with heat and humidity is something the British are, on the whole, not very good at. Noel Coward implied that our habit of going out in the midday sun was mad, and we can all picture the streets of the City of London thronged with gents in three-piece suits even during spells of tropical heat. What happens, then, if we fail to cope?

In great heat, with the temperature in excess of, say, 35°C (95°F), the effects of both high and low humidity can be very dramatic. Outdoor physical effort is at best extremely uncomfortable, at worst downright dangerous. This is as true of vigorous sporting activity as it is of labour on a building site or in a garden.

Excessive humidity means that our bodies have to work very hard to keep our internal temperature stable: we perspire freely, try to minimize physical exertion, wear loose light-coloured clothing, seek shade, drink water, and attempt to create air movement by fanning ourselves. In extreme cases these mechanisms may be insufficient, so our body temperature rises and heatstroke follows. Untreated, it will lead rapidly to death. This combination of heat and high humidity is most strongly marked around the coastal fringes of the Arabian Gulf – for example in Dubai, Abu Dhabi and Bahrain. Elsewhere, long periods of dangerously humid weather occur in the Gulf States of the USA, in Far Eastern cities such as Singapore and Bangkok, along the coasts of both East and West Africa, and in Queensland. But, as our climate gets progressively warmer, even our corner of the planet is at risk, as we found to our cost during the unprecedented heatwave of August 2003 when 20,000–30,000 people died across western and central Europe. British figures were never clearly stated, but it is probable that between 1,000 and 2,000 people died from heat stress that month in the UK.

Exceptionally high temperature allied to exceptionally low humidity is just as perilous. Again we perspire rapidly to try to cool our bodies although we are less aware of it because the perspiration quickly evaporates in the very dry air. Unless we maintain body fluid and body salts by frequent drinks of a suitable nature we soon become dehydrated and disorientated, sunstroke follows, and this too may eventually be fatal. The combination of heat and low humidity can be found in any desert region within the tropics and this is one of the two reasons (the other is the absence of water supply) why most of the Sahara Desert and the Australian interior are effectively uninhabitable.

Both heatstroke and sunstroke occur more quickly when the head is bare, because a substantial quantity of the Sun's heat energy is absorbed through the scalp. Thus loose-fitting light-coloured headgear is strongly advised for anyone who expects to be out in the sun for any length of time. Alcoholic drinks are not a good idea: alcohol absorbs water, our bodies become less efficient at maintaining a stable internal temperature, and this accelerates the onset of both heatstroke and sunstroke.

Weather experts often combine information about temperature and humidity in an attempt to give the public a better idea of expected levels of discomfort. Such information may be given in the form of a 'temperature–humidity index' or THI, sometimes more simply called a 'comfort index'. Some forecasters in the USA and UK have been known to use the allegedly more user-friendly name 'humisery index'.

The intensity of sunlight is much greater in tropical regions compared to the UK during the middle of the day, since the sun is almost overhead and therefore shines through a smaller thickness of atmosphere than it does when it is lower in the sky. Sunburn therefore occurs much more rapidly than it does at home – within 15 minutes if the atmosphere is clear of haze or dust – so a strong sun-block cream should always be applied before spending any time outdoors. The purpose of this is to minimize the amount of ultraviolet radiation which gets through to our skin. All sunlight contains ultraviolet light, although most of it is filtered out by the ozone layer in the upper atmosphere. However, some of it gets through, and prolonged exposure to it is harmful because it accelerates the wrinkling and ageing of the skin, it causes sunburn, and it is responsible for some melanomas (skin cancers). In sunny climates colonized by white-skinned peoples, such as in California and Australia, between 25 and 50 per cent of all cancer deaths among Caucasians are from melanomas induced by ultraviolet radiation. It may also cause a range of eye complaints which in turn can lead to persistent headaches and neuralgia.

So you can see that spending time in a hot sunny climate is fraught with all sorts of dangers. Enjoy your holiday!

Lightning and thunder

Flickering lightning at the horizon, distant thunder grumbles long and low, the first huge drops of rain the size of half-crowns splatter on the kitchen roof below. Unexpectedly a much closer flash fills one's field of vision, momentarily blinding; the arm goes up in a reflex but forlorn effort to shield one's face; immediately a great peal of thunder like a giant heap of bricks tumbling to the ground rends the silence. The hairs on the back of the neck stand on end in response to that familiar marriage of fear and excitement. The good old British summer has done it again: a few days of hot sunny weather, the growing humidity, the ugly threatening clouds gathering ominously throughout the oppressive afternoon and evening, and the inevitable storms follow after dark.

Watching night-time thunderstorms through the bedroom window was a secret childhood hobby of mine, one that I have never completely grown out of. So, if the poor old dear up the road who hides in the space under the stairs when storms are brewing is known to her doctor as a 'brontophobe' then I guess that I, along with many other meteorologists, must be some sort of a 'brontophile'. I think I should keep quiet about it.

'It's God, shifting his furniture around,' some of us were told as tiny tots, when we cowered, uncomprehending, at the awesome violence of an overhead thunderstorm. A few years older, and pestered grown-ups would give us the benefit of their lack of knowledge: 'Thunderstorms happen when two clouds collide,' they offered. Oh, that it were so simple. In times past, pagan gods were invoked to explain the power and fury of the full-blooded thunderstorm, and the ancient Norse deity, Thor, usually depicted wielding a hammer, was the god of thunder.

The harnessing of electricity for use in the factory and the home may be a relatively recent development, but electricity was known to the Ancient Greeks. *Electrum* is the Latin word for amber, and comes directly from the Greek word *elektron*, and as long ago as the sixth century BC ancient philosophers knew that amber, when rubbed, would attract small pieces of cloth. They failed, however, to make the link between this rather unexciting form of 'static' electricity and the drama of the thunderstorm. It was not until the American scientist, Benjamin Franklin, performed his famous experiment of flying a kite beneath a thundercloud that it was proved beyond reasonable doubt that lightning was an electrical manifestation.

Contrary to popular belief, he did not succeed in attracting a lightning stroke to his kite – he would probably not have survived such an event – but he did collect sufficient electrical charge down the rain-soaked line to initiate a spark between his hand and a metal key which was attached to the rope.

The cumulonimbus is king of the clouds. This is the giant towering thunder-head which under the right conditions boils furiously upwards to a height of 15 kilometres (about ten miles) or more. Along with thunder and lightning, these dramatic clouds bring torrential downpours, explosive hail-storms, savage squalls of wind, and on occasion even tornadoes. And all this comes out of thin air – literally – thanks to the interaction of warmth from the sun and invisible water vapour in the atmosphere.

Thunderclouds are characterized by vertically moving air currents called updraughts and downdraughts. The updraughts carry sizeable raindrops from the bottom to the top of the cloud where the temperature may be as low as –50°C (–58°F), by which time the raindrop will have frozen into a small hailstone. The hailstone may remain in the cloud for some time, depending on the strength and staying power of the updraught, and it is believed that collisions between small hailstones and tiny ice crystals create an electrical charge. The charge on the hailstones is negative, while that on the ice crystals is positive. The migration of the ice crystals to the top of the cloud, and the congregation of the larger hail-pellets in the lower half of the cloud result in a separation of electrical charge in the cumulonimbus cloud, negative at the bottom and positive at the top. The surface of the Earth and objects on it such as trees and buildings also carry an electrical charge. When a sufficient difference in charge has built up, electrical discharges will occur, both between the cloud and the ground and also inside the cloud. These electrical discharges are what we know as lightning.

All lightning is 'forked' lightning, but if the discharge has occurred within a dense cloud mass, or at a great distance, all you may see is a diffuse reflection of the lightning stroke on the water droplets of the cloud or on other clouds nearby. This is generally termed sheet lightning or summer lightning. A bolt of lightning carries a massive electrical current, estimated at anything up to 50,000 amps, which produces around 100 megawatts of power per metre, which in turn heats the air along the lightning stroke to a temperature of 30,000°C (54,000°F) or more.

A clap of thunder is the necessary response of the atmosphere to a flash of lightning. When a flash occurs, the air in the lightning channel heats up by thousands of degrees in a tiny fraction of a second. The heated air expands explosively, sending out a violent pressure wave which is followed by a succession of expansions and compressions due to the fact that the air is, in effect, an elastic medium. These are what we hear as thunder.

The lightning flash may take place in a fraction of a second, but the sound of thunder usually lasts several seconds, and under the right atmospheric conditions can continue for the best part of a minute. This is because the speed at which sound travels is relatively slow, taking about three seconds to cross one kilometre, or five seconds to travel a mile. Thus a slow count up to five between seeing the lightning and hearing the thunder – a habit most children learn at their mother's knee – tells us that the electrical discharge occurred roughly a mile away; a count of ten indicates two miles, and so on. The thunder continues to rumble for several seconds after the initial crash because it takes different times for the sound to reach us from different parts of the lightning channel.

Many summer thunderstorms experienced in the UK occur fairly high in the atmosphere: meteorologists call this *medium-level instability*. The clouds which produce these storms may be around three kilometres (two miles) above the ground, and the lightning discharges occur within the cloud and are therefore often obscured. As a consequence, we observe sheet lightning rather than forked lightning. On such an occasion, even an overhead flash is three kilometres away, and the far end of the lightning channel may be at more than twice that distance. At such distances the higher-frequency elements of the thunder are damped down, and we are just left with a low-frequency 'growl' which starts some ten seconds after the flash and may carry on rumbling for a further ten seconds or more.

Winter thunderstorms, rare in lowland Britain, take place in cumulonimbus clouds which sit low in the atmosphere. Indeed, the base of the cloud may be only 400 metres (1,300 feet) above the ground. The initial crash of the associated thunder is therefore surprisingly loud, so much so that I've known professional meteorologists taken aback by it.

What to do in a thunderstorm

Some people hide away in the cupboard under the stairs. And from our earliest years we are taught never to shelter under trees or to touch metal window-frames. Thunderstorms – especially lightning – are feared by more people in this country than any other sort of weather.

But there is some good news if you are worried about being struck by lightning. The statistics show that the risk of being killed by a bolt of lightning diminished sharply during the course of the twentieth century, and the latest figures indicate that this trend is continuing into the present century. In 1900 the chance of any one person suffering a fatal injury in a single year during a thunderstorm was 2 million to one. By 1950 that figure had changed to 4 million to one, and by the year 2000 the odds had lengthened to 15 million to one. In Victorian times roughly an equal number of men and women were killed by lightning, but now the death rate is heavily tilted towards males, with men six times more at risk than women.

A politician and his spin doctors could, no doubt, weave wonderful stories around such statistics: it is a vindication of government policy to reduce the risk of adverse weather to the general populace, or it illustrates how successfully our schools have driven home the message of not standing under trees during storms, or maybe the successful installation of lightning conductors on buildings and other tall structures, championed by previous Labour/Conservative (delete as necessary) governments, is now producing excellent results.

Before any MP reading this rushes for pen and paper to seek more information, I have to reveal that there is a much simpler explanation for these rather startling figures. There has, in fact, been no significant change in the frequency of thunderstorms in the UK during the last 200 years. But there has been a marked change in human activity during this period. The number of people working in the open air has decreased dramatically, in particular farmers and agricultural labourers who used to spend most of their time in the fields, especially during the spring and summer when storms are at their most frequent. Nowadays fewer women than men work outdoors, and in any case a sizeable proportion of lightning injuries in recent decades has been borne by sportsmen – golfers, cricketers and footballers in particular.

If you are still worried by that one in 15 million chance, then you can do no better than follow these lightning safety rules, based on those issued by the National Weather Service in the USA:

1. Stay indoors; keep away from open doors and windows, and from metal furniture such as radiators, stoves, metal pipes and stainless-steel sinks. Do not use plug-in electrical appliances or the telephone.

2. Do not take washing off the clothesline.
3. If you are working outdoors, move away from fences, telephone or power lines, or other metal structures. Stop tractor work, and keep clear of other farm equipment in the open.
4. If you are indulging in leisure pursuits, get out of the water and off small boats. Don't use metallic implements such as fishing rods and golf clubs. Remove shoes with metal studs and toecaps. Rubber-soled shoes afford a surprising degree of protection.
5. If you are in your car stay there; if you can reach it quickly, climb inside and shut the door. Motor cars offer excellent lightning protection.
6. If you are caught in the open, seek shelter in any nearby building. If there is none, your best protection is in a cave, in a cleft in the mountainside or cliff-face, or in a ditch.
7. When there is no shelter, avoid the highest object in the area. Crouch down, keeping knees and feet together. If you feel a growing electrical charge – your hair stands on end or your skin tingles – a flash of lightning is imminent and you must drop to the ground immediately.

One other statistic should be mentioned here. Fewer than half the people struck by lightning are killed. Most receive a severe electrical shock and many suffer serious burns, but recovery from lightning strikes is usually complete except for possible impairment of sight or hearing. Someone who appears to have been killed can sometimes be revived by prompt mouth-to-mouth resuscitation and heart massage. In a group struck by lightning, as on a football or cricket field, those who appear lifeless should be treated first as those who show signs of life will almost certainly recover spontaneously.

The idea that lightning never strikes twice is a fallacy. The Eiffel Tower, for example, is struck 25 times per year on average. The most outrageous human lightning conductor was a park ranger from Virginia, USA, called Roy Sullivan, who was struck seven times between 1942 and 1977. He subsequently committed suicide, presumably unwilling to face an eighth strike.

Missiles from the sky

Have you ever been hit on the back of the head by a hailstone? It is not funny. Even small ones are quite painful. Every summer, several severe hailstorms are reported from somewhere or other in the UK, with stones around 2.5 centimetres (up to an inch) across. That is fairly big by British standards: big enough to strip trees of their leaves, to pulverize young crops and garden flowers, and on such occasions there are often instances of birds being killed in large numbers as well.

The only time I have experienced hailstones of that size was during a thunderstorm on the evening of Friday 7 June 1996. I spent five minutes or so desperately trying to cover my car with blankets and old carpet fragments in an attempt to prevent damage to the paintwork, and I was hit several times. Not an experience I'd care to repeat. Down the road, where the stones were bigger, there were hundreds of dead sparrows by the roadside, in adjacent gardens greenhouses were smashed to pieces, and car windscreens were chipped.

A large hailstone is an offensive weapon. We can calculate approximately the characteristics of such a missile to give us some idea of its damage potential. Any object falling under the influence of gravity will reach a maximum speed when the weight of the object is balanced by the resistance of the air. That resistance depends on the object's density, shape, and how smooth or rough it is. The maximum speed is called the 'terminal velocity'. For a snowflake it is about one metre per second (two miles per hour), for a tiny drizzle droplet it is 3 m/sec (6 mph), and for a typical raindrop 6 or 7 m/sec

(12–16 mph). Our 2.5-centimetre (one-inch) hailstone weighs approximately seven grams (a quarter of an ounce) and travels at a speed of about 20 m/sec (45 mph).

Serious falls of hail are nearly always associated with thunderstorms. Raindrops are carried by powerful updraughts high into the upper reaches of thunder clouds (cumulonimbus) where the temperature may be as low as –35°C (–30°F). Here they freeze, transforming into small hailstones. Eventually the updraught runs out of steam, allowing the new hail pellets to fall; some of them reach the ground, but some will be picked up anew by the updraught and will collect a film of moisture from the impact of raindrops, this moisture again freezing as the pellet reaches sub-zero parts of the cloud. This cycle of rising and falling is sometimes repeated several times, the hailstones growing larger with each cycle. Cut a large hailstone in half and you will see several rings of lighter-coloured ice, rather like the rings of an onion, each one representing a single round trip in the thundercloud.

As the cumulonimbus reaches maturity the updraughts weaken and the millions of hailstones still held aloft become too heavy to stay there. Within a matter of minutes the cloud ejects most of its icy contents. Viewed from the ground the effect is very dramatic. The storm suddenly breaks, a deluge of hailstones clatter to earth, and although the temperature minutes before may have been above 32°C (90°F) the ground is rapidly covered with ice. Most hailstorms last less than ten minutes and the worst-hit area is usually very small. Experts talk about a 'hail streak' because damage from a typical hailstorm affects a long, narrow zone, perhaps 400 metres (a quarter of a mile) wide by three kilometres (five miles) long. This elongation is a consequence of the storm's movement.

In Britain one of the most damaging storms on record struck Essex, Hertfordshire and Bedfordshire on the afternoon of Midsummer Day – 24 June – in 1897, and it ruined some of the events marking Queen Victoria's diamond jubilee. Poultry and game-birds were wiped out in their thousands. Some livestock were seriously injured and had to be put down, and several people suffered cuts and bruises. One huge, flattened hailstone picked up at Writtle, near Chelmsford, measured 10 cm by 7.5 cm by 2.5 cm (four inches by three inches by one inch), and weighed 140 grams (five ounces).

The costliest storm hit Munich in Germany on 12 July 1984. Hospitals dealt with hundreds of casualties, but no lives were lost. Insurance losses were put at 750 million pounds sterling, largely resulting from damage to motor cars.

On rare occasions several hailstones fuse together into a huge, irregularly shaped chunk of ice. The largest authenticated such stone was picked up after a hailstorm at Coffeyville, Kansas, USA, on 3 September 1970. This monster measured 20 centimetres (eight inches) across and weighed in at 0.75 kilograms (27 ounces). Such a stone would reach a speed of almost 50 m/sec (around 110 mph) – that's five times as heavy as, and 20 mph faster than, a cricket ball propelled by Mr Andrew Flintoff. Imagine being at the wrong end of that. Now that's what I call a *terminal* velocity . . .

Chapter 5

The four seasons: autumn

Introduction

> It's a long, long time from May to December,
> And the days grow short when you reach September;
> When the autumn weather turns the leaves to flame,
> One hasn't got time for the waiting game.
>
> Kurt Weill ('September Song')

The division of the year into four seasons divided by the equinoxes and solstices was first proposed by an ancient Rhodean astronomer called Geminus. He lived in a part of Europe where there can be no confusion between the astronomical seasons and the natural seasons, because the rains, the wind, the sunshine and the heat all arrive pretty much on cue every year. Nor should there be any confusion in the United Kingdom even though our seasons are less well marked than they are in the Mediterranean. A few people like to insist that the astronomers' seasons are somehow the 'official' seasons, but the simple fact is that in Britain there is no official way of dividing up the year.

Autumn in the Middle Ages was normally taken to be August, September and October, as these were the months of the corn harvest, the gathering in of fruit and nuts, and the putting down of stores of food for the winter. The Church told its congregations *Autumnat Bartholomaeus; dat Clemens hiemem* – autumn lasts from the feast day of St Bartholomew on 24 August until 22 November, with winter taking over on the 23rd which is St Clement's Day.

The analysis of *natural seasons* carried out by Professor Lamb revealed consistently autumnal weather types between the second week of September and the third week of November, and he proposed that on this basis *autumn* should run from 10 September to 19 November. For the convenience of climatological book-keeping, however, we still take the months of September, October and November as the autumn quarter, and, Geminus notwithstanding, this is the nearest thing to an 'official' autumn that we have in the UK.

September in the twentieth century

> September dries up ditches
> Or breaks down bridges.
>
> Anon.

One of my favourite September weather sayings did not in fact originate in Britain, but it is certainly appropriate to our climate at this time of the year. For most of the twentieth century, September has been a markedly drier month than either August or October, although the changing climate of the 1990s has confounded that particular principle as it has confounded so many others. This change is not down to any excessive wetness of September as the end of the century approached; rather it was a consequence of drier Augusts and Octobers.

Over the years there have certainly been a large number of very wet Septembers; these are usually markedly 'Atlantic' months when westerly and south-westerly winds blow practically throughout, and vigorous depressions advance from the west across the British Isles at frequent intervals. The Atlantic

Ocean is at its warmest during early September, so air-masses which reach us from that direction are both warm and moist – warm air has the capacity to support more moisture than cold air has – and are able to deposit large quantities of rain. The downpours are occasionally supplemented by the remnants of tropical hurricanes which get caught up in the south-westerly flow from time to time. Cloudy skies and blustery winds keep daytime temperatures subdued, typically between 14 and 18°C (57–64°F), but nights can be surprisingly warm during these spells of Atlantic weather. Rarely does a full-blown northerly develop bringing yet colder air directly from the Arctic. These unsettled Septembers, both rainswept and windswept, are clearly autumnal months.

By contrast, there have also been many Septembers which one could quite reasonably describe as summery. Very dry months with a total rainfall, averaged over England and Wales, of 30 millimetres (1.2 inches) or less occur more frequently in September than August – twelve times against six during the twentieth century. Such Septembers are dominated by high-pressure systems, bringing cool misty mornings, warm sunny afternoons, light breezes and a notable absence of rain and thunder. September warmth is rarely extreme, thanks to the waning power of the sun, the shorter day-length, and the chilly mornings, but very occasionally, say once every 20 years, Britain really does get to swelter in September.

Of course many Septembers – perhaps most – fall between two stools, with spells of wet and windy weather alternating with fine, warm interludes. In this way autumn advances in fits and starts during a typical year.

Decade	*Mean temperature*	*Rainfall*	*Sunshine*	*Thunderstorms*
1880–89	13.0°C	85 millimetres	120 hours	–
1890–99	13.5°C	71 millimetres	148 hours	1.7 days
1900–09	13.1°C	56 millimetres	149 hours	1.3 days
1910–19	13.0°C	64 millimetres	146 hours	1.5 days
1920–29	13.2°C	75 millimetres	147 hours	1.1 days
1930–39	13.7°C	80 millimetres	131 hours	2.1 days
1940–49	14.0°C	68 millimetres	133 hours	1.1 days
1950–59	13.5°C	88 millimetres	134 hours	1.5 days
1960–69	13.5°C	90 millimetres	133 hours	1.1 days
1970–79	13.4°C	76 millimetres	144 hours	1.2 days
1980–89	13.8°C	76 millimetres	136 hours	1.4 days
1990–99	13.9°C	81 millimetres	133 hours	1.6 days
2000–04	14.3°C	67 millimetres	144 hours	0.8 days

The analysis of September temperatures during the twentieth century shows a roughly 50-year oscillation, with cold months predominating up to 1930 and again between 1950 and 1980, while the warmest months occurred in the 1940s and 1990s. The marked warming trend in the last 30 years of the century found in practically every other month of the year is certainly present in the September record, although the 1990s just failed to be the warmest decade of all. The first years of the twenty-first century suggest that the warming may be accelerating slightly.

Septembers were largely dry during the first two decades of the century, but they became significantly wetter from about 1920 onwards; those of the 1950s and 1960s recorded on average 50 per cent more rain than those of the 1900s and 1910s. No obvious link between temperature and rainfall can be made, although there is a slight tendency for the warmer decades to be wetter, and the colder decades to be drier. The frequency of thunderstorms is markedly lower than during

July and August, averaging one to two days per month throughout the period.

The sunshine record shows a curious pattern. The first three decades of the twentieth century were consistently sunny (as indeed the 1890s had been), while the rest of the century was generally cloudier, with some 10 per cent less sunshine. There was no clear association between sunshine amount and either temperature or rainfall, although again there is a barely detectable tendency for the sunnier decades to be dry but relatively cool. The dry/cool relationship is characteristic of the winter half-year, while the dry/sunny one is typical of the summer half, illustrating the transitional character of September.

Heatwave of the century

Undoubtedly the most dramatic autumn heatwave of the twentieth century occurred early in September 1906, when the temperatures soared above 32°C (90°F) on the first three days of the month (as well as on 31 August). The hottest day of the spell was 2 September when the temperature climbed to 34°C (93°F) over a large part of England, peaking at 35.6°C (96°F) at Bawtry, near Doncaster. That is just 2.5 degrees short of the all-time record for any month of the year, yet it happened barely three weeks before the autumn equinox.

Flood of the century

South-east England is the driest corner of the country, on average, but during the middle of September 1968 that was turned on its head. A broad swath of country stretching from the New Forest to the Thames Estuary had over 75 millimetres (three inches) of rain in 24 hours, and at Tilbury and Stifford, both in south Essex, the downpour deposited just over 200 millimetres (close to eight inches). Put another way, that was three months' worth of rain in just one day. Floodwaters covered 250 square kilometres (almost 100 square miles) and 25,000 homes were inundated, chiefly in Surrey, around the suburban towns of Esher, Molesey, Walton, Weybridge and Guildford. London was inaccessible from the south and south-west for nearly two days.

Gale of the century

September's strongest winds, sometimes called equinoctial gales, frequently arrive courtesy of a former hurricane or tropical storm. During September 1961 two ex-hurricanes, Betsy and Debbie, swept past Britain's western seaboard. Debbie was the more violent, and the wind gusted to 98 knots (113 mph) in the north of Ireland and to 91 knots (105 mph) in the Hebrides. Shipping was badly disrupted, thousands of trees uprooted, electricity and telephone services cut off, and hundreds of homes damaged. Northern Ireland was the most seriously affected part of the UK, but the Irish Republic suffered widespread gale damage too.

Cold snap of the century

The only occasion that snow was authentically observed in southern England in September was on the 20th in 1919. Sleet showers were reported in the London area, and snow fell sufficiently heavily to leave a covering on the ground over the West Country moors, the Welsh hills and the higher ground of the Midlands. On Snowdon the snow was 15 centimetres (six inches) deep and persisted for a week. There were sharp frosts in most parts of the country, killing off summer flowers and vegetables.

October in the twentieth century

Whatever the weather, October is always the month when the full ripeness of the year finally turns to decay. Fruit fall and rot if not harvested, the trees take on their autumn colours, and the railway companies dust off their book of excuses for leaves on the line. During the 1940s, 1950s and 1960s, climatologists became aware of a marked tendency towards warmer and drier Octobers, and indeed a number of Octobers up to 1972 resembled a late-summer month rather than the traditional descent towards winter. This trend was abruptly checked with the exceptionally cold October of 1974, the coldest since 1905, and although there have since been strong hints of a resumption of the warming process, the recovery has not been entirely convincing.

On the surface, then, the strong warming in the 1980s and 1990s evinced in most other months of the year appears to be absent in October. A closer look at the weather patterns during the last few decades explains why. The strong mid-century warming was due to an increase in the frequency of southerly winds and a complementary decrease in the frequency of northerlies. Since then the respective frequencies of southerly and northerly winds have returned to early twentieth-century levels, and temperatures have fluctuated widely, giving us notably cold Octobers such as 1992 and 1993, and outstandingly warm ones such as 1995. The new century has brought record-breaking warmth in 2001, and a rare spell of severe frosts in 2003. The table opposite shows clearly the remarkable warmth of the 1960s, and the dryness of the 1970s.

The main feature of October temperature, apart from the exceptional nature of the 1960s, is the step change which appeared to take place around 1940. Until then the warmest decade was colder than the coolest in the later period, and the mean temperature between 1890 and 1939 of 9.6°C (49.3°F) contrasted with 10.4°C (50.7°F) for 1940–2000. Put another way, that is equivalent to a delay of eight days in the decline towards winter.

Changes in October rainfall are less clear-cut. The driest

Decade	*Mean temperature*	*Rainfall*	*Sunshine*	*Thunderstorms*
1880–89	8.9°C	116 millimetres	90 hours	–
1890–99	8.9°C	103 millimetres	92 hours	0.9 days
1900–09	9.9°C	108 millimetres	95 hours	1.0 days
1910–19	9.3°C	94 millimetres	98 hours	1.2 days
1920–29	10.0°C	96 millimetres	116 hours	0.6 days
1930–39	9.7°C	100 millimetres	99 hours	0.8 days
1940–49	10.4°C	90 millimetres	103 hours	0.7 days
1950–59	10.2°C	75 millimetres	98 hours	0.5 days
1960–69	10.9°C	90 millimetres	94 hours	0.5 days
1970-79	10.4°C	67 millimetres	101 hours	0.8 days
1980–89	10.3°C	109 millimetres	107 hours	0.2 days
1990–99	10.4°C	86 millimetres	113 hours	0.6 days
2000–04	10.7°C	135 millimetres	100 hours	0.1 days

decade was the 1970s with a mean rainfall of 67 millimetres (2.64 inches), which was immediately followed by the wettest decade, the 1980s, with 109 millimetres (4.29 inches). If one ignores that blip in the 1980s, there was a tendency towards less rain in the second half of the century, and three of the driest decades occurred after 1950. However, the early years of the twenty-first century have been characterized by a dramatic return of very wet Octobers. Thunderstorm frequency in October has shown a general decline during the last 100 years or so.

The blip in the sunshine pattern occurred early in the century. The sunniest decade was the 1920s with 116 hours, but that apart there was a gradual improvement in sunshine hours as the century progressed, and this brightening-up process has been strongly marked in the last 40 years, with the average of 94 hours in the 1960s contrasting with 113 hours in the 1990s. Once again, though, the turn of the century brought yet another change of direction.

Heatwave of the century

It was not easy to choose between the brief but intense hot spell of early October 1985 when a new all-time record for the month was established, and the long spells of exceptional warmth in 1959 and 1921 when long hot summers lingered well into autumn. In the end I picked none of these, settling instead for the five-day heatwave in October 1908 which was all the more welcome because it followed a poor summer. It was also unusual for the location of the highest temperatures – in Yorkshire, Cumbria, north Wales, the west and north of Scotland, and parts of Ireland. Hottest days were the 3rd and 4th when the mercury soared to 28.3°C (83°F) at Whitby on the north Yorkshire coast.

Flood of the century

The West Country floods of October 1960 brought untold misery to thousands of people, especially in Taunton and Exeter. This was a particularly unpleasant month because severe flooding recurred at roughly weekly intervals from early October to mid-November, so that just as people had cleared away the mess of one inundation, another one arrived. In some parts of Devon some 375 millimetres (15 inches) of rain fell during the month, four times the average amount.

Gale of the century

It could only be Michael Fish's 'hurricane'. The one which, legend has it, was not supposed to happen. It struck on the night of 15–16 October 1987, and the worst-hit area was south-east of a line drawn from Dorset to The Wash. Weather historians believe that this was the most violent and most destructive windstorm to hit southern England since 1703. Eighteen people died – if it had struck during daylight hours the death toll could have been over one hundred – millions of trees were uprooted, and insurance losses were calculated at £2,500 million converted to today's prices. The strongest gust in the UK that day was 100 knots (115 mph) at Shoreham in Sussex.

Cold snap of the century

October 1974 was the coldest for almost 70 years, and not a single day during the month saw the temperature reach the seasonal average. The 7th was a particularly nasty day, with a bitter north wind, and wet snow fell in Essex and Hertfordshire for almost an hour around breakfast time. The day's maximum temperature here was just 7°C (45°F). There have been colder snaps in October, but this one happened so early in the month, and followed that long run of summery autumns in the 1960s and early '70s, so it really did come out of the blue.

November in the twentieth century

November has two faces, meteorologically speaking. One is grim and gloomy: day after day of dull dismal skies, a thick mist obscuring the familiar horizons, autumn leaves lying about, damp and inert. The other is fierce and threatening: wild windswept days, driving rain beating against the windows, the wind soughing in the trees, the dead leaves swirling vigorously or crackling drily underfoot. Both of these faces reveal important characteristics of November's personality, but there are one or two other traits as well.

Occasionally November surprises, and delivers us a spell of mellow autumn sunshine. Or a sudden surge of air from the Arctic sends us a succession of crisp sunny days with cloudless skies and sharp night frosts. But rarely does November don winter garb and dump a foot of snow on our doorstep. The following table shows the decade-by-decade changes in temperature, rainfall, sunshine, and the frequency of snow lying on the ground, since 1880, enabling us to place recent fluctuations in November weather into some sort of historical context.

Decade	*Mean temperature*	*Rainfall*	*Sunshine*	*Snow cover*
1880–89	6.3°C	98 millimetres	55 hours	–
1890–99	6.7°C	86 millimetres	54 hours	0.7 days
1900–09	6.1°C	76 millimetres	56 hours	1.3 days
1910–19	5.7°C	92 millimetres	65 hours	2.5 days
1920–29	5.8°C	97 millimetres	63 hours	1.5 days
1930–39	6.8°C	103 millimetres	50 hours	0.3 days
1940–49	6.7°C	97 millimetres	54 hours	1.1 days
1950–59	6.7°C	100 millimetres	52 hours	1.1 days
1960–69	6.1°C	98 millimetres	55 hours	2.5 days
1970–79	6.7°C	98 millimetres	73 hours	1.2 days
1980–89	6.8°C	88 millimetres	66 hours	2.0 days
1990–99	7.1°C	96 millimetres	63 hours	1.4 days
2000–04	7.8°C	113 millimetres	61 hours	0.4 days

November temperatures during the twentieth century reflect the erratic warming trend which is apparent in most months of the year. There was a hiccup in the 1960s when we experienced several decidedly cold and snowy Novembers thanks to a marked but temporary increase in the frequency of northerly winds; even so, the '60s were not as cold as the 1910s and '20s. The general upward trend appeared to accelerate in the 1990s which was by far the warmest decade in 340 years of records, and temperatures have continued to rise rapidly during the first years of the new century. The 1900–29 average of 5.9°C (42.6°F) contrasts with a figure for 1970–99 of 6.9°C (44.4°F). This is the equivalent of a 12-day delay in the onset of winter. November rainfall has been remarkably consistent since 1910, but the first decade of the twentieth century was much drier than any other. Unlike most other months of the year, there is no trend evident in the decade-by-decade averages.

In complete contrast, the sunshine figures seem to be all over the place. There is a broad fluctuation from sunny

Novembers in the 1910s and 1920s to dull months between the 1930s and early 1960s, then a marked improvement during the 1970s and 1980s, followed by another decline since then. Broadly speaking, these wide variations reflect changes in the relative frequency of sunny north-westerly and northerly winds on the one hand, and gloomy south-westerlies and southerlies on the other.

The snowfall statistics are interesting. The snowiest decades were the cold ones of the 1910s and 1960s when northerly winds were most prevalent. But the warmer years since 1980 have produced a surprisingly large number of snowy days, which indicates that there continue to be northerly outbreaks in spite of the underlying warming trend.

Freeze of the century

The first anniversary of the Armistice was held on a clear, brisk morning on 11 November 1919. This day marked the beginning of a phenomenal spell of Arctic weather rarely equalled even in January or February. The temperature dropped to –23.3°C (–10°F) at Braemar in Aberdeenshire on the morning of the 14th, and this remains the November record by a wide margin. Even in midwinter it has only been surpassed five times since. A snowstorm swept the country on the 11th and 12th, dumping 20 centimetres (8 inches) on Edinburgh, 30 centimetres (12 inches) on Dartmoor, and 42 centimetres (17 inches) on Balmoral.

Flood of the century

November 1929 was one of the wettest of all time, and the daily downpours were heaviest in Wales and the West Country, leading to prolonged and widespread flooding. The wettest day was the 11th, when 211 millimetres (8.3 inches) of rain fell at Mardy reservoir in Glamorgan. The floods were worst in the Rhondda Valley and in Pontypridd where the swiftly flowing waters carried coal slurry and other debris into hundreds of houses.

Gale of the century

Right at the beginning of the twentieth century, on 12 November 1901, a sudden and unpredicted gale precipitated one of the greatest human disasters before the Great War. The storm centre tracked across southern Ireland, Wales, the Midlands and Lincolnshire, and the high winds were confined to those areas south of this track. Over 200 people drowned as several ships went down in the Channel and the South-west Approaches; the wind averaged 55–65 knots (63–75 mph), with strongest gusts close to 90 knots (just over 100 mph), building massive seas around our southern and south-western coasts. This storm may have contained the remnants of a Caribbean hurricane.

Warm spell of the century

Not normally known for producing heatwaves, there have been three short spells during the Novembers of the twentieth century which saw the temperature climb to 21°C (70°F) or above – in 1938, 1946 and 1997. The highest temperature of all occurred on 6 November 1946, when the mercury soared to 21.7°C (71°F) at Prestatyn in north-east Wales. The 20°C (68°F) threshold was crossed in Manchester, Leeds, Edinburgh, several places along the north Welsh coast, and as far north as Nairn in north-east Scotland.

Harvest home

> There was three Kings into the east,
> Three Kings both great and high;
> And they hae sworn a solemn oath:
> John Barleycorn should die.
> Robert Burns ('John Barleycorn')

Usually just a euphemism for alcoholic drink these days, John Barleycorn was also the personification of the harvest, the spirit of the fields. Pre-Christian communities in the British Isles celebrated the completion of the harvest around the autumn equinox, and this celebration continued after the Christianization of our islands, in a festival known as Harvest Home. One of the climaxes of this essentially pagan holiday was the mock sacrifice of John Barleycorn, a large wickerwork figure containing the last sheaf to be harvested, and sometimes dressed up rather like a scarecrow. Presumably one purpose of this ritual was to ensure the benevolence of the gods during the following year.

For centuries Harvest Home ran side by side with the Harvest Festivals celebrated in churches up and down the land, although the rural clergy tended to turn a blind eye to the often raucous and sometimes sleazy goings-on in the fields. The church holiday became Michaelmas, and was fixed to 29 September. This is fairly close to the present position of the autumn equinox which falls on or around 22 September (in fact it can occur on the 21st, 22nd or 23rd, depending on where we are in the leap-year cycle). However, we should remember that the switch from the Julian to the Gregorian calendar in 1752 corrected a slippage of 11 days; this slippage had arisen because all century years (1200, 1300, 1400, 1500, etc.) were regarded as leap years in the earlier calendar whereas in the present system only century years divisible by 400 are. Thus in the earlier part of the eighteenth century the equinox was around 11 September, in the fifteenth century it was normally on 13 September, and in the thirteenth century it fell on or about the 15th.

Whether Christian or pagan, these September holidays were important, not just as a celebration of a successful harvest but also to mark the autumn equinox – that pivotal point in the cycle of the year when the hours of darkness exceed the hours of daylight for the first time since March. Summer really is over. Winter looms.

By our medieval forebears September was actually regarded as the middle month of autumn. In the rural society which Britain was before 1800, 'autumn' was almost synonymous with 'harvest', not just the harvesting of corn but also the gathering of fruit, berries and nuts. For them autumn began with Lammas (originally 'loaf mass') Day, 1 August, and ended at All Hallows' Eve, 31 October. If there is any doubt as to the importance of Michaelmas to our ancestors in the cycle of the seasons, we need only refer to the ancient volumes of old country weather lore. There are dozens of sayings referring to this date, some predicting the next year's harvest, many forecasting what the winter will be like, and all of them in one way or another indicating the onward march of the seasons. As one old French saying has it:

> *La Saint-Michel, la chaleur rentre aux cieux*
> [At Michaelmas, the warmth returns to the heavens].

Forecasting fog formation

'. . . and with all the moisture about after the recent heavy rain, mist and fog patches will probably form later in the night.' How many times have you heard statements like that in television and radio weather forecasts?

There is a myth in some forecasting offices around the country that a wet autumn or winter inevitably leads to a higher incidence of fog, assuming other factors are favourable, than is the case during a dry one. The argument is that excess water lying in the fields is somehow picked up by the air above and then transformed into fog. Broadly speaking, fog formation does not work like that.

However, there is one circumstance where excess surface

moisture does contribute to fog development, and that is on occasions of 'advection fog' or 'thaw fog'. In Britain this typically occurs after a lengthy cold spell, when a chill east wind is replaced by a warmer but very moist southerly one. The cold ground, often waterlogged or covered with melting snow, cools the air flowing across it, and because cool air cannot support as much water vapour as warm air, some of the moisture condenses in the form of water droplets – that is, fog. But even in this circumstance the fog formation is a consequence of the cooling effect of the surface water rather than a transfer of moisture from the ground to the air.

It can even be argued that the fog that forms on calm, cloudless nights – meteorologists call this 'radiation fog' – is *less* likely to occur when the ground is saturated, especially after several days of very mild weather. Water has a much higher thermal capacity than solid materials such as soil and asphalt, so at night it cools down more slowly. This means that air in contact with the Earth's surface will also cool down more slowly over waterlogged ground than it does over dry ground. The slower it cools, the longer it takes to reach saturation point, and the later the fog will form.

Over the years we can find a great many examples to support this hypothesis. For example, it is certainly true that there were several foggy dawns during August 1976 – the culmination of the great drought of that year – and the same was true during the middle fortnight of September 2003 following that summer's record-breaking heatwave. By contrast, fog was not particularly widespread or persistent during the week-long quiet spell in mid-December 2000 when large tracts of Yorkshire, the Midlands and southern England lay under floodwaters following that autumn's repeated deluges.

The Horncastle flood

It is now approaching half a century since the appalling weather disaster which struck the small Lincolnshire market town of Horncastle, some 25 kilometres (15 miles) east of Lincoln.

A large low-pressure system settled over the British Isles during the first week of October 1960, bringing widespread heavy downpours, but there were also some sunny breaks, and when the sun shone it became warm and very humid. On the 6th the temperature reached 21°C (69°F) at Maldon in Essex, and on the morning of Friday 7th some brief gleams of sunshine again saw the temperature climb quickly. But the rising temperature triggered the rapid build-up of thunder clouds, and storms broke out at several places in southern, eastern and central England during the afternoon. These were essentially summer thunderstorms even though it was October, with serious localized cloudbursts and intense electrical activity.

Between midday and 6 pm a total of 183.9 millimetres (7.24 inches) of rain was measured by the local observer at Horncastle, a Mr F. J. Harvey. He reported that most of the rain fell between 2.30 and 4 pm. That is the equivalent of over three months' worth of rain in just 90 minutes. It should also be added that it is highly unlikely that the rain-gauge in the town was precisely at the point of highest rainfall so it is probable that the peak fall was even larger. Nevertheless, this remains the heaviest fall inside a day ever recorded in Lincolnshire. (The UK record is 279.4 millimetres [11 inches] in 15 hours at Martinstown, Dorset, on 18 July 1955.)

The high street was turned into a raging torrent which carried away caravans and cars; large quantities of stock from flooded shops together with household furniture were also swept away to the lower part of the town. A local man was found drowned the following morning, no more than 100 metres from his home. Over 100 houses were flooded, and losses were estimated at £100,000 which translates to £1.5 million at today's prices. It was almost a year before several of the worst-damaged houses and shops were again habitable.

Seven kilometres (four miles) away at Revesby reservoir, the total fall from the deluge was 161.5 millimetres (6.36 inches). The duty technician wrote in his logbook that the water level in the reservoir rose from 2.5 metres to 3.5 metres (eight feet to eleven feet) during the afternoon, while the boiler and filter houses were flooded to a depth of three metres (ten feet). The nearby River Waring overtopped its banks, flooding hundreds of hectares of farmland.

The Horncastle storm was by no means the only torrential downpour during that early part of October 1960, although it was undoubtedly the most intense. Devon and Dorset were badly hit on the 2nd, with 60 millimetres (2.36 inches) of rain in two hours at Rousdon, near Lyme Regis; on the 8th 103 millimetres (4.05 inches) fell at Luxborough in Somerset; and the following day 87 millimetres (3.42 inches) was measured at Danby Lodge in north Yorkshire.

The floods of October 1903

The floods of autumn and winter 2000–01 will remain in most people's memories for a very long time. During October 2000 alone, more than twice the normal amount of rain fell over much of the UK, but it was not, by some distance, the wettest October of all. That occurred almost 100 years earlier. Averaged over England and Wales, rainfall in October 2000 was 188 millimetres (7.40 inches), which is the equivalent of 230 per cent of the long-term average, but in October 1903 it was 218 millimetres or 8.59 inches (268 per cent). Not only was it the wettest October on record, it was the wettest month of any name since useful rainfall records began over 300 years ago.

The greatest monthly catch was on the flanks of Ullscarf above Borrowdale in Cumbria, where 813 millimetres (32 inches) of rain fell, but the largest deviations from normal rainfall were found in the middle Thames Valley and also on Tyneside where Tynemouth reported more than three times the average amount. Rain fell, little or much, on at least 28 days over much of the UK, and quite large regions reported rain on every single day of the month.

Serious flooding was reported from all corners of the country, and towards the month's end approximately 15,000 hectares of farmland were under water, chiefly along the courses of the Yorkshire Ouse, the Severn, the Trent, the Great Ouse and the Thames. Fewer householders were flooded out compared with the events of 2000–01, but in 1903 the habit of building housing estates on flood plains lay well in the future. However, the floods caused several major structural failures to buildings, and bridges, notably at Risca in Monmouthshire where four arches of a 12-metre (40-foot) high viaduct crossing the Ebbw River collapsed.

Farming was badly hit. In southern and most midland counties the harvest had been safely gathered in by the end of September, but from Staffordshire and Derbyshire northwards crops were abandoned in the fields. James Smith wrote from Crathes, 25 kilometres (15 miles) west of Aberdeen, '. . . crops were in a deplorable condition, the grain was soft and not very marketable . . . the turnip crop was poor and potatoes showed signs of disease', and the Reverend J. B. Duncan at nearby Alford wrote, '. . . harvest was very backward, a good deal of corn being still out at the end'.

Rain was not the only problem in October 1903. The vigorous weather also brought hailstorms, severe gales, thunder and lightning, and even a couple of tornadoes. On the 5th, heavy hail damaged glasshouses in the western and southern suburbs of London, and the next day violent winds disrupted transport and uprooted trees. On the 25th, tornadoes caused much damage in the Wareham district of Dorset and near Banbury in Oxfordshire, many homes in southern England were struck by lightning, and floodwaters demolished dozens of bridges in Cornwall and Devon.

The big difference between 2000 and 1903 was that the more recent year brought further heavy falls of rain in November and early December, resulting in repeated episodes of flooding in all the major river systems of England and Wales. In 1903, by contrast, the waterlogged October was followed by much drier weather during November and December, which allowed the floodwaters to subside steadily.

The Great Storm of 1703

It is almost two decades since Michael Fish's hurricane blasted across southern England and northern France. At the time it was widely described as the most intense and destructive storm to sweep this part of the world since the Great Storm of 1703, and the subsequent detailed analysis of the 1987 storm provided no evidence to change that view.

Generally speaking, our knowledge of weather events before the advent of systematic instrumental records in the mid-1800s is sketchy. But we know a lot about the 1703 gale thanks to the diligent scholarship of Daniel Defoe. More famous perhaps for *Robinson Crusoe*, Defoe produced in 1704 a slim volume entitled *The Storm or, A Collection of the Most Remarkable Casualties and Disasters which happen'd in the late Dreadful Tempest both by Sea and Land.* A great lumbering title it may be, but the contents present us with a careful historical account deliberately designed by the author to set down for posterity all the information he could gather.

We are also indebted to the twentieth-century climatologist, Professor Hubert Lamb, who raided a variety of historical archives including ships' logs to construct approximate daily weather charts for the fortnight centred on the date of the storm itself – 26–27 November in the Julian calendar (7–8 December in the calendar we use today).

Lamb's reconstruction shows a small but exceedingly intense secondary depression travelling almost 1,500 kilometres (900 miles) from the South-west Approaches to Denmark in 24 hours, the centre of the low crossing mid-Wales, the north Midlands, Lincolnshire, and the central North Sea, where the barometric pressure at the eye of the storm dropped below 950 millibars. The strongest winds, as is normal with quick-moving secondaries like this, were on its southern flank, including all of southern England, the English Channel, the southern North Sea, and the Low Countries. Lamb estimated sustained winds of 70–80 knots (about 80–90 mph) with gusts to 108 knots (125 mph).

This extraordinarily violent gale was the culmination of a prolonged period of very windy weather. As a consequence, many seagoing vessels had already sought shelter in harbours and estuaries when the tempest struck. Nevertheless, an estimated 30 per cent of the English merchant fleet was lost, and over 7,500 seamen perished including 1,500 men of the Royal Navy. On the Thames, between Shadwell and Limehouse, Defoe describes how 700 vessels were driven together and left in a hopeless tangle, and at London Bridge some 60 lighters and barges were forced up against the piers although the bridge itself survived with only minor damage. Another totem of English naval and technological power, the five-year-old Eddystone lighthouse, was swept away by the pounding waves, taking its designer, Henry Winstanley, with it.

On land many homes were ruined and over 400 windmills were wrecked – some were simply blown down, but others burned to the ground because the wooden machinery overheated as the sails were forced to rotate far faster than they were supposed to. The Bishop of Bath and Wells was killed when a chimney stack fell through the roof into his bedchamber. Curiously, his wife survived without a scratch!

Unexpected snow

Rockall apart, they are geographically the most remote part of the British Isles, although 'remote' is not a word that most people would use to describe them. They are also closer to the European continent than any other of our islands. They are warmer and sunnier, on average, than anywhere in Great Britain or Ireland. They are, of course, the Channel Islands.

Jersey is the largest island of the group, the nearest to France, and the furthest from the English mainland; it is almost 160 kilometres (100 miles) from Portland Bill, but barely 16 kilometres (ten miles) from the Normandy coast. Their distance from Britain, their proximity to France, their maritime location, and their southerly latitude contribute to the Channel Islands' warm and sunny but rather moist climate. These factors also mean that the day-to-day weather here can be very different from the weather on the British mainland, and this is particularly true of Jersey.

There are occasions when Jersey is the only British island to enjoy a European heatwave, but there are others when England and Wales are warm and sunny but the Channel Islands are fog-bound. A destructive gale in October 1964 caused millions of pounds worth of damage in Jersey and Guernsey but did not touch the English south coast. And on 31 October 1985, Jersey airport recorded the lowest October maximum temperature at a lowland site in the British Isles – just 3°C (37°F) – for over 40 years, courtesy of a bitter east wind blowing straight off the French mainland.

That 16-kilometre hop from France means that severe winter weather in Europe sometimes arrives with scarcely any modification; by contrast the British Isles in general are protected from continental extremes by the relative warmth of the surrounding seas. It is for this reason that the weather can play outrageous tricks on Channel Islanders, as in early November 1980 when another easterly wind blew without a break for over a week. At times the wind backed north-easterly, streaming down the length of the English Channel and bringing frequent showers of rain or hail. But from late on the 4th until the early hours of the 7th the wind came from due east, pulling in air which had originated over northern Russia several days before. Jersey, as one might expect, was more severely affected than the other islands. It snowed almost continuously for 48 hours with the temperature close to freezing point, and on the 6th snow lay five centimetres (two inches) deep at Jersey airport and eight centimetres (over three inches) deep in St Helier. The day's maximum temperature on the 5th at the airport was just 0.1°C (32.2°F). In over 150 years of weather recording, substantial snow lying on the ground in the Channel Islands has never been noted so early in the season.

The wintry weather in early November 1980 was not confined to Jersey and Guernsey. Several centimetres of snow also lay for a time over the hills of southern England, especially in Kent, Surrey and Sussex, although here it melted far more quickly than it did on Jersey, and at no point did the snow cover extend to sea level as it did at St Helier. Later that same month, when a northerly air-stream blowing directly from the Arctic brought widespread snowfalls to the British mainland, the Channel Islands escaped thanks to that broad and relatively warm expanse of English Channel that the north wind had to cross before reaching them.

November warmth

The stunning weather of early November 2003 is now but a memory, but it is one that will stay with us a long time. Highest temperatures on Thursday afternoon, 6 November, were 18.8°C (65.8°F) at Northolt and Kensington, and 18.4°C (65.1°F) at Heathrow, while on Friday 7 November the warmest weather migrated westward and northward, giving 20.2°C (68.4°F) at Lochcarron and 19.5°C (67.1°F) at Aultbea, both in Wester Ross, 19.6°C (67.3°F) at Llanbedr in Merionethshire, 19.1°C (66.4°F) at Broadford on Skye, and 18.8°C (65.8°F) at Kinloss in Morayshire. The statistics tell only part of the story, though, and most non-meteorological people will remember the crystal-clear blue skies, the uninterrupted sunshine, the exceptional autumn colours, and the peculiarly warm breeze while being quite ignorant of the numbers.

Exceptional warmth in November in the UK is almost always imported, delivered from subtropical latitudes by a strong south-westerly or southerly wind. In the lee of high ground the warmth is accentuated by the compression of air descending the mountain slopes, a mechanism known as the 'föhn effect', after the warm and very dry föhn wind of the Swiss and Austrian Alps. November sunshine is relatively feeble and low in the sky, so the warmth of the sun contributes little to temperature levels on these rare warm days.

The highest November temperature ever recorded in the UK under standard conditions was 21.7°C (71°F), measured at Prestatyn, Flintshire, on 4 November 1946. Nearby Hawarden Bridge reported 21.1°C (70°F), and although maxima of 18–20°C (64–68°F) were observed widely elsewhere it is clear that on this occasion the föhn effect provided sufficient added impetus to give the record to the north-eastern corner of Wales. Similarly, Edinburgh's 20.6°C (69°F) and Dublin's 20.0°C (68°F) on the same date are the highest November temperatures ever recorded in Scotland and the Irish Republic respectively.

England's highest for the month happened on 5 November 1938 when 21.1°C (70°F) was recorded at several sites in East Anglia and the south-east including London, Chelmsford and

Cambridge. On this occasion a stiff south-westerly breeze was accompanied in eastern counties by several hours of bright sunshine which lifted temperatures by 2–3 degC (3–5 degF) when compared with cloudier regions further west and north.

The third example of extreme November warmth was a recent occurrence, in 1997, when new date-records were established on 15–18 November inclusive. Cloudy skies and a strong south-easterly wind meant that the föhn effect was very much in evidence, and this was emphasized by the fact that highest readings were obtained along the coast of north Wales. Indeed, the maximum on the 17th of 20.7°C (69.3°F) at Aber, seven kilometres (four miles) east of Bangor, occurred at 7 pm – nearly three hours after sunset.

Autumn colours

Autumn used to be my least favourite season. Then one year, 1980 I think, while working in the Middle East, I returned home for some leave towards the end of October. The drive home from Heathrow was a revelation: the brilliant autumn colours set against a deep blue sky, the early-morning sunshine slanting brightly through the trees. Why had I never noticed this before?

It was, perhaps, an illustration of the truism that you have to be deprived of something to appreciate it properly. After months in the desert, the sky white, the sun glaring, the ubiquitous sand a dreary dun, nearly all colour bled out of the environment, the sudden change to late October in England could hardly have provided a greater contrast. Ever since, I have looked forward to those few weeks when the autumn colours are at their best.

Where do these autumn hues come from? The fact is that a variety of pigments, including greens, yellows and reds, are present during the spring and summer, but the greens are so dominant that they completely mask the other colours. This is the result of chlorophyll production, which is essential to the conversion of carbon dioxide and water into the nutrients necessary for the trees to grow. In autumn the process ceases, the chlorophyll in the leaves breaks down, the green pigments disappear, and the other colours become dominant. The change is triggered by the progressive drop in light-levels, and in particular by the shortening day-length. The temperature of a particular season may accelerate or delay the process by a few weeks, and a warm autumn, as in 1994 or 2001, may see the autumn colours last until mid- or late November.

Some seasons are less good, especially when early frosts and strong winds remove the leaves more quickly, and when an absence of blue sky and sunshine diminishes the contrasts of colour around us. In any case, our autumn tints in Britain are rather bland compared with those found in Canada and the USA. They, of course, have developed a tourist industry around what they call the 'fall foliage season'; New England is probably the best region to visit, but the shortness of the season guarantees high prices.

The best colours occur during cool sunny autumns with little rain and light winds, and without severe frosts. The best recent years in the UK were probably 1996 and 2003. The leaves are still alive at this stage, but a hard frost will cause the supply of nutrients to the foliage to cease and the leaves will drop almost immediately. The displays are more dramatic in North America, not because of climatic differences, but because of the different trees which are native to the area. Many maples produce stunning scarlets and crimsons, as do some American oaks, hornbeam, some laurels, sorrel, sassafras, tupelo and sumach. White ash and wych-hazel, meanwhile, are known for their purple and maroon tints. In the UK our most startling colours are seen on introduced species, but the golds and coppers of sycamore, oak and beech are nevertheless stunning on a sunny November day.

Chapter 6

The four seasons: winter

Introduction

> In the bleak mid-winter,
> Frosty wind made moan,
> Earth stood hard as iron,
> Water like a stone.
>
> Christina Rossetti

Born in 1830, Rossetti knew London when winters really were winters. Europe was very gradually coming out of the so-called Little Ice Age, but during the nineteenth century roughly 20 per cent of English winters were in the 'very cold' or 'severe' category. During the twentieth century there were just seven, and the last of those was in 1978–79 – over a quarter of a century ago. What price earth as hard as iron these days?

One thing that does not change, however, is the length of the day, and midwinter days are just as short and dark as they ever were. The winter solstice – the shortest day – falls on or around 21 December, and this marks the first day of winter in the astronomers' seasonal scheme; the last day of winter comes just before the spring equinox, on 21 March give or take a day. The medieval ecclesiastical calendar tells us that winter runs from St Clement's Day which is 23 November, and gives way to spring on St Peter's feast day which is 22 February – *Dat Clemens hiemem; dat Petrus ver.*

When Britain was a predominantly agricultural society, the general perception was that November, December and January were the deadest months of the year, thus comprising winter, whereas February brought the first green shoots, and drying winds often provided an opportunity to get out on the land. In modern times, December, January and February are considered to be the winter months, and these months also comprise the winter quarter for climatological purposes.

Professor Lamb's attempt to distinguish the *natural seasons* of the year resulted in two seasons straddling the coldest months. His analysis showed that *early winter* runs from about 20 November until 19 January, while *late winter and early spring* takes over between 20 January and 31 March.

December in the twentieth century

The tradition of white Christmases probably owes more to the change from the Julian to the Gregorian Calendar in the middle of the eighteenth century than it does to anything or anyone else. Before that change occurred in 1752, when the year 'slipped' by 11 days, Christmas fell in what would now be the first week in January. The frequency of snow is significantly higher then compared with late December. But in no period in the last 250 years has the Christmas holiday been a particularly snowy one, and those familiar Christmas card snow scenes have always been very much the exception rather than the rule.

We can extend the generalization to December as a whole. Most of the severest winters in history did their worst during January and February, and occasionally even March. Snowy Decembers are rare: there were just two during the twentieth century, in 1950 and 1981. Over lowland Britain there is an average of just two days per month with snow falling, and three days per month with snow covering the ground. Much more often December appears to be a late-autumn month, characterized by frequent rain and gales, occasional brief colder interludes with fog and frost, and distinguishable from November only by the shortness of daylight. Once upon a time, December was an even gloomier and dirtier month than it is now, but the clean-air legislation of the 1950s and 1960s delivered a marked improvement in the sunshine records of all our towns and cities.

The following table describes the changes in temperature, rainfall, sunshine, and the frequency of snow, since 1880, and it highlights some surprisingly large contrasts, especially between the 1960s and 1970s:

Decade	*Mean temperature*	*Rainfall*	*Sunshine*	*Snow cover*
1880–89	3.2°C	89 millimetres	58 hours	–
1890–99	3.7°C	89 millimetres	56 hours	1.5 days
1900–09	4.3°C	93 millimetres	55 hours	1.2 days
1910–19	5.1°C	127 millimetres	65 hours	1.9 days
1920–29	4.5°C	98 millimetres	63 hours	2.2 days
1930–39	4.4°C	85 millimetres	50 hours	3.3 days
1940–49	4.8°C	83 millimetres	54 hours	0.9 days
1950–59	4.9°C	95 millimetres	52 hours	3.0 days
1960–69	3.5°C	97 millimetres	55 hours	3.1 days
1970–79	5.3°C	93 millimetres	73 hours	1.6 days
1980–89	4.2°C	99 millimetres	66 hours	2.9 days
1990–99	4.6°C	101 millimetres	63 hours	1.3 days
2000–04	5.0°C	99 millimetres	53 hours	1.0 days

A decade-by-decade analysis of December temperature shows the same warming process during the twentieth century that we have seen in the other months of the year, but it has been a very erratic warming. There was a huge jump from a mean temperature of 3.5°C in the 1960s to 5.3°C in the 1970s, followed by a sharp drop back to 4.2°C in the 1980s, and a slow recovery since then. The acceleration of the warming trend in the 1990s evident in several other months of the year has simply not been present in December. In fact, the 1910s, '40s, '50s and '70s were all warmer than the '90s.

There is no obvious trend in rainfall, with most decades averaging between 90 and 100 millimetres (3.6–4 inches). However, the 1910s stand out as an exceptionally wet decade, averaging 127 millimetres (5.0 inches), while the 1930s and 1940s were relatively dry at 83–85 millimetres (about 3.3 inches) each. Sunshine averages 60 hours over the whole century, but there was quite a well-marked improvement from just 50 hours in the 1930s to 73 hours in the 1970s, followed by a slight easing back during the 1980s and 1990s. The December of 2001 was sunnier than any during the previous century.

An examination of the snowfall statistics shows a correlation between the number of days with snow cover and the mean decadal temperature, which is pretty much what one would expect. However, the Decembers of the 1940s were curiously snow-free, while the 1950s had approximately the same temperature but more than three times as much snow. In 53 of the last 100 Decembers there was no day with a general snow cover, while in 1950 there were 17 such days and in 1981 there were 22.

Freeze of the century

The severe weather of December 1981 crept across the nation with hardly any warning at all. There was no sudden surge of bitter northerlies blowing directly from the Arctic, or piercing easterlies from Siberia. During the 6th and 7th rain was gradually replaced by sleet and then by snow, and there was hardly a breath of wind. Some 20 centimetres (eight inches) of snow lay across much of southern England by the evening of the 8th, and a further 15–20 centimetres (six to eight inches) fell on top of that on Friday the 11th. Subsequently, under clear skies and with a deep snow cover temperatures dropped to record-breaking levels. During the early hours of the 13th much of England together with east Wales and southern Scotland lay in a deep freeze at –10°C (14°F) or below, and some places were very much colder than that. At RAF Shawbury in Shropshire the mercury plunged to –25.2°C (–13.4°F), the lowest ever recorded in England in December.

Snowstorm of the century

The Christmas blizzard of 1927 was one of the most destructive and disruptive snowstorms of the entire twentieth century. Rain fell for much of Christmas Day over central and southern England, but a gradual drop in temperature resulted in the rain being replaced progressively by snow during Christmas afternoon and the storm raged all night and for most of Boxing Day too. Level snow was 50–60 centimetres (around two feet) deep over the Chiltern Hills and the Downs of Berkshire, north Hampshire and Wiltshire, but drifts six to ten metres high were widely reported. Many main roads were not reopened until the New Year, but then a rapid thaw accompanied by heavy rain led to serious flooding, notably along the River Thames.

Flood of the century

A week of repeated downpours over Christmas 1978 led to widespread flooding. Between the 26th and 28th some 245 millimetres (9.6 inches) of rain fell at Silent Valley in the Mourne Mountains of Northern Ireland. Conditions were worst in and around Belfast, on the Ouse at York, and in the Welsh Valleys. The flood was followed by a dramatic drop in temperature during the 29th and 30th, and in Yorkshire in particular flooded fields turned into skating rinks by New Year's Eve.

Gale of the century

30 December 1900 saw a ferocious depression race across Ireland and northern England, with severe gales on its southern flank. Gusts of wind between 80 and 90 knots (92–103 mph) were logged at exposed sites around the coasts of Wales and southern England, thousands of trees were uprooted, and many buildings were badly damaged. Several ships foundered in the Channel and the South-west Approaches, and the death toll from the gale exceeded 200. The gale was accompanied by a prolonged downpour which dropped up to 90 millimetres (3.6 inches) of rain in the Midlands.

Warm spell of the century

Achnashellach, on the railway line from Inverness to Kyle of Lochalsh deep in the north-west highlands of Scotland, is hardly the sort of place one would expect to a hold a temperature record for warmth. But on 2 December 1948 the mercury here climbed to 18.3°C (65°F), the highest ever logged in the UK in December. A mild southerly wind forced to climb over the Scottish highlands was warmed by compression – the so-called föhn effect – as it descended into Glen Carron, the sun shone and the temperature soared. Several other weather-recording sites in north-west Scotland and also along the southern shore of the Moray Firth recorded highs of 15 or 16°C (round about 60°F) the same day.

January in the twentieth century

They've stolen our winters. In most parts of Britain January has been the coldest month of the year since months were invented – to be pedantic some of our western and south-western coastal fringes are colder in February though only by a fraction – but it is many years since we last had a truly wintry January. We have to go back to 1987 for the last time we experienced a cold snap lasting more than two weeks during the first month of the year, and to 1979 to find a January which could reasonably be described as 'severe'.

Oddly, although it is marginally warmer, there have been many more severe Februarys than Januarys; this is a reflection of the way long spells of easterly winds tend to become established relatively late in the winter season, say in mid or late January, and then may persist right through February, bringing frigid air from north-eastern Europe. Cold Januarys usually comprise a succession of wintry snaps separated by rather milder interludes – 1979 was one of these – with the result that mean monthly temperatures are not excessively low. The one exception to this pattern was January 1963 when frost and snow persisted for ten weeks.

The decadal statistics for January describe a month which has changed dramatically since the late 1930s. During the middle of the century it was a cold month that delivered lengthy wintry episodes quite regularly, but by the 1990s it was a full two degrees warmer, and snow was almost relegated to a thing of the past.

Decade	*Mean temperature*	*Rainfall*	*Sunshine*	*Snow cover*
1880–89	3.6°C	67 millimetres	42 hours	–
1890–99	3.3°C	73 millimetres	52 hours	7 days
1900–09	3.9°C	73 millimetres	55 hours	2 days
1910–19	3.9°C	85 millimetres	43 hours	2 days
1920–29	4.7°C	98 millimetres	50 hours	3 days
1930–39	4.5°C	100 millimetres	45 hours	3 days
1940–49	2.7°C	96 millimetres	46 hours	8 days
1950–59	3.4°C	80 millimetres	51 hours	5 days
1960–69	3.4°C	83 millimetres	47 hours	7 days
1970–79	4.1°C	98 millimetres	40 hours	5 days
1980–89	3.8°C	91 millimetres	62 hours	5 days
1990–99	4.7°C	100 millimetres	52 hours	2 days
2000–05	4.7°C	85 millimetres	63 hours	2 days

These figures show that Januarys grew progressively warmer during the first 30 years of the century, reaching a notable peak during the 1920s and 1930s. This was followed by a dramatic drop in the 1940s which comprised the coldest decade since the 1840s – four of the six coldest Januarys of the century were those of 1940, '41, '42 and '45, with 1947 just a fraction warmer. (The famous snowy winter of 1947 did not really get going until 22 January and it lasted till the middle of March.) The other two severe Januarys were in 1963 and 1979. The cold interlude persisted throughout the 1950s and 1960s, but an erratic warming trend has been apparent since about 1970. The 1920s and 1990s brought the warmest Januarys in the entire Central England Temperature record which began 345 years ago, and the years since 2000 have maintained that remarkable level of warmth.

The colder periods also tended to be slightly drier while the warmer ones were notably wet: the six cold decades averaged 85 millimetres (3.36 inches) of rain, whereas the four warm decades averaged 99 millimetres (3.91 inches). There was no significant association between sunshine and the other elements, and the main feature of the sunshine record was the remarkable brightness of the 1980s which averaged one-third more sun than the preceding eight decades, and half as much sunshine again as the 1970s. As one might have expected, there was a strong link between temperature and snow cover: snow on the ground was four times as frequent during the cold 1940s and 1960s compared with the warm 1990s.

Statistics rarely reveal the big memorable or newsworthy events. Here are my choices for January:

Disaster of the century

There is no question that the North Sea floods of 31 January 1953 top the list, since this was undoubtedly the worst wholly natural disaster to hit Western Europe during any month for

250 years. (The killer smog of December 1952 took more lives, but the deadly component of that meteorological event was man-made pollution.) A violent northerly gale coincided with a high spring tide, piling billions of gallons of water into the increasingly constricted geography of the southern North Sea. The death toll was 307 in the UK (but more than 1,600 in the Netherlands); a total of 750 square kilometres (300 square miles) of eastern England was flooded, with 32,000 people made homeless; and insurance losses on both sides of the North Sea were around £20,000 million (at present day prices).

Snowstorm of the century

Between 26 and 29 January 1940, a combined snowstorm and ice-storm swept much of England, Wales and Northern Ireland, completely paralysing transport systems and leaving the UK militarily very vulnerable. For this very reason weather reports and forecasts were regarded as top secret throughout the war, and newspaper reports of this storm did not appear until almost three weeks later. In the Midlands, level snow lay 45 centimetres (18 inches) deep with massive drifts, while a broad swath extending from west Sussex and Hampshire across Wessex, the Thames Valley, the south-west Midlands to mid- and north Wales experienced a prolonged fall of freezing rain, coating everything in clear ice over an inch thick.

Gale of the century

The so-called 'Burns' Day Storm' of 25 January 1990 swept the whole of the UK south of Glasgow and Edinburgh with gusts locally as high as 94 knots (108 mph). Widespread structural damage occurred, 47 people died, and high-sided vehicles blown over littered the fringes of most motorways. This severe gale affected an area eight times larger than did the

October 1987 storm and killed almost three times as many people, and yet it is almost entirely forgotten by the popular media. Why? Probably because it was accurately predicted.

Freeze-up of the century

Top place must go to January 1963, when the greater part of England and Wales was snow-covered throughout the month, frost occurred on every night, and at some places in Wiltshire the temperature did not once reach 2°C (36°F). Surprisingly, it was also a dry and sunny month especially in western districts, and there were only two heavy falls of snow, on the 3rd and the 19th. That winter was the coldest, averaged nationally, since that of 1739–40, and the 'Big Freeze', as it was called at the time, lasted from 22 December until 4 March. Throughout the winter the warmest parts of the UK were regularly to be found in the Hebrides.

Heatwave of the century

January is hardly a month when one might expect high temperatures, but a rare combination of meteorological circumstances led to a bizarre heatwave – perhaps 'mildwave' would be a more accurate description – on 10 January 1971. The temperature climbed to 18.3°C (65°F) at Aber on the north coast of Wales between Bangor and Llandudno, and readings in excess of 16°C (61°F) were widely recorded to the lee of high ground in Wales, northern England and Scotland. This was another instance of the föhn effect which occurs when a warm and moist air-mass, usually delivered by a broad south-westerly airflow, is forced to rise over ranges of hills and mountains where it drops much of its moisture, then descends the lee slope as a dry, clear and very warm breeze. On this occasion the warm wind was a southerly one, hence the highest readings were observed on the northern flank of Snowdonia.

February in the twentieth century

'The cold grows stronger as the days grow longer', according to the old country saying. The days are shortest and the sun is lowest at the winter solstice, at or about 21 December, but the coldest weather in the UK usually arrives in January or February. This is because our climate is partly controlled by the seas which surround the British Isles, and there is an awful lot of thermal energy locked up in such large bodies of water. Thus the sea-surface temperature drops slowly during the autumn and winter, finally reaching its lowest in late February or early March, then rises gradually during the spring and summer to reach its highest in early September.

Averaged over a long period of years, January is fractionally colder than February over most of Britain, but there is very little in it. Let us look at it another way. During the last century, January was the coldest month in 43 years, February in 28 years, December in 21, March in 5 and November in 3. One striking statistic is that, since 1900, intensely cold Februarys have outnumbered similarly wintry Januarys five to three. Even so, it should be emphasized that really severe Februarys are very much the exception rather than the rule, and that is why they stick in the collective memory for such a long time. Each generation has a childhood memory which men and women of a certain age can throw at you without provocation . . . 'Ah, but we don't have real winters any more, not like 1979.' Or 1963, or 1947 or 1929. The 1895 generation has long gone.

Statistics for the twentieth century illustrate how February changed as the decades passed, and the truly exceptional nature of the 1990s stands out clearly:

Decade	*Mean temperature*	*Rainfall*	*Sunshine*	*Snow cover*
1880–89	3.9°C	68 millimetres	51 hours	–
1890–99	4.0°C	63 millimetres	61 hours	6 days
1900–09	3.6°C	63 millimetres	65 hours	4 days
1910–19	4.4°C	77 millimetres	61 hours	4 days
1920–29	4.6°C	72 millimetres	63 hours	2 days
1930–39	4.2°C	63 millimetres	67 hours	2 days
1940–49	3.7°C	60 millimetres	61 hours	7 days
1950–59	3.5°C	73 millimetres	61 hours	5 days
1960–69	3.6°C	61 millimetres	57 hours	5 days
1970–79	3.9°C	74 millimetres	61 hours	3 days
1980–89	3.5°C	56 millimetres	74 hours	6 days
1990–99	5.2°C	73 millimetres	74 hours	3 days
2000–05	5.4°C	81 millimetres	83 hours	1 day

These figures reveal no clear-cut trend in February temperature during the twentieth century. If anything, there was a slight decline from the 1910s to the 1980s, but then we had a dramatic upward lurch between the 1980s and 1990s. This huge jump appears to be a long way outside previous decade-to-decade changes, not just in February but in any month of the year. In fact a search through the records for earlier centuries showed that the period 1990–99 produced the warmest Februarys in the 345-year-long record. It is worth emphasizing here that the Januarys of the 1990s were also the warmest in the entire record. The first few years of the twenty-first century indicate that the unprecedented level of warmth continues, although there has been only a small additional rise in temperature. The coldest individual Februarys were scattered among the decades – in 1917, 1929, 1942, 1947, 1956, 1963 and 1986. By contrast, of the nine warmest Februarys of the century, four occurred in the 1990s – 1990, 1995, 1997 and 1998.

At first sight there is no strong trend in February rainfall either, but a closer examination reveals that the warmer

Februarys were on the whole wetter, while the colder ones were drier. The five warmest decades averaged 72 millimetres (2.85 inches) of rain, while the five coldest averaged 62 millimetres (2.45 inches). Sunshine amounts fluctuated around the average until the 1970s, but since 1980 February appears to have become appreciably brighter, with an average of something like 20 per cent more sunshine than earlier decades. The 1940s stands out as the snowiest decade, thanks to the severe winters of 1940, 1941, 1942 and especially 1947 when most of Britain was snow-covered throughout February, but the 1980s were not far behind.

The years 2000–05 comprise too short a period to draw any real conclusions, but it is worth drawing attention to the fact that this period has been warmer, wetter, sunnier and less snowy than any other.

The Februarys of the twentieth century have thrown up their fair share of dramatic and historic weather events, and even a few which are just plain peculiar. Here are some of the most remarkable:

Disaster of the century

Without doubt the most sustained period of high winds in the historical record swept the UK between 25 January and 27 February 1990. Depression after depression travelled across the country, bringing a series of severe gales with gusts widely in excess of 70 knots (80 mph) on each occasion. The worst of these gales, on the 26th and 27th, coincided with a high spring tide. North-westerly winds gusting to 86 knots (99 mph) drove a storm surge into the coasts of north Wales, the Wirral and Lancashire, breaching the sea-defences in several places. A major disaster overwhelmed Towyn, between Rhyl and Abergele, when parts of the sea wall collapsed and seawater rushed inland, inundating over a thousand homes. Some 4,000 people were evacuated, and many never returned and their badly damaged houses had to be demolished. There was a hidden catastrophe, too: many of those who lost their homes were elderly people who had retired to the seaside and, in straitened circumstances, had chosen not to take out insurance policies.

Gale of the century

A similar windstorm struck the country on the same dates, 26 and 27 February, in 1903. Worst hit were northern England and Wales, with sustained winds of 45 to 55 knots (52–63 mph), and peak gusts in excess of 80 knots (92 mph). The wind was strong enough to derail a passenger train travelling across the Leven viaduct near Ulverston in north Lancashire. Ten carriages were overturned, and there were many casualties. If anything, the wind was even stronger and the damage more widespread in Ireland where it was generally regarded as the severest gale since 'The Night of the Big Wind' in January 1839. Dublin suffered terribly: there was extensive damage to roofs and chimneys in the city, and over 3,000 trees were uprooted in Phoenix Park alone.

Snowstorm of the century

A three-day snowstorm hit much of the British Isles between 23 and 26 February 1933. It was as remarkable for its unexpectedness as for its severity, coming in the midst of a sequence of mild and snow-free winters. A deepening depression stalled off south-west Ireland, and the associated belt of heavy rain became stationary over England, Wales and Ireland, the rain readily giving way to wet snow as it engaged the bitterly cold easterly airflow which had covered the country for some days before. Little snow fell south of a line from Essex to Bristol, but 60–90 centimetres (two to three feet) of snow covered much of the rest of the country, completely disrupting transport.

Freeze-up of the century

The winter of 1962–63 may have been the coldest of the century, but that of 1947 was the snowiest, and February that year was characterized by relentless easterly winds and frequent heavy snowfalls. Over much of the Midlands and northern England the temperature failed to climb above 2°C (36°F) during the entire month, and the ground was deeply snow-covered throughout. At Ushaw, near Durham, for instance, snow lay 30 centimetres (a foot) deep on the 1st, and 70 centimetres (28 inches) deep on the 28th. February 1947 was also remarkable for the lack of sunshine over England and Wales: many places including London experienced 21 consecutive days without a solitary sighting of the sun.

Heatwave of the century

The warmth of the 1990s reached a climax in the middle of February 1998. Between the 8th and the 26th the temperature regularly climbed to between 12 and 15°C (54–59°F), reaching 19.1°C (66.4°F) at Tivington in Somerset on the 14th, and 18.1°C (64.6°F) at Prestatyn in Flintshire on the 15th. A report of 19.6°C (67.3°F) at Worcester on the 13th, found in several books of record, is now discounted. Worcester apart, these are among the highest temperatures ever authentically recorded in Britain in February, and are all the more remarkable for having occurred in the middle rather than at the end of the month. Over most of England and Wales the warmth was accompanied by blue skies, bright sunshine and a moderate breeze, and the whole month was one of the driest Februarys of the century in many southern and eastern districts.

Christmas cuckoos

Not long ago, as the end of the year approached, I had a telephone call from a reporter at a radio station in London, and she asked me the question that, come December, weather people dread. 'When was the last time we had a white Christmas?' The first cuckoo, I thought.

For some reason my usual amiability faltered. 'Who is "we"?' I asked. 'What do you mean by a white Christmas? *Where* are you talking about? And why do you want to know?' Nonplussed, she paused, then she tried again. 'But surely, it's ever such a long time since we had a white Christmas. I can't remember the last time. It must be ages.'

'Yes,' I said. 'It's a *very* long time . . . in fact it's very nearly 12 months. It last snowed at Christmas, well, last Christmas. I don't suppose you can remember that far back.' Somehow sensing she would rather the conversation ended there, I wasn't going to let her go so easily. But I was mildly surprised how patiently she listened to my little diatribe about how the concept of a white Christmas had been perverted by the bookies, how the statistics are all geared to their requirements, and how they won't pay out unless flakes of snow are reported falling on the roof of the London Weather Centre. Snow on the ground simply does not count. Britain could be buried under half a metre of snow on Christmas morning, but if it all fell on Christmas Eve the bookmakers would hang on to their dosh. As for the eccentrics, con-artists, and other kinds of self-styled long-range forecasters who claimed to be able to predict a Christmas snowfall three weeks before the event . . .

It was at this point I heard the click of the journalist's phone. I don't know about her, but I felt a lot better.

On average, snow or sleet falls on 25 December in London during one year in twelve, on average. Strangely, given the warming trend in our climate, there have been several examples since 1993 although none of them produced a widespread snow cover. In northern Britain, Christmas 1995 was very white with deep snow in northern and eastern Scotland and north-east England, and the Shetland Islands were without power following two tremendous snowstorms in the preceding week. Christmas 2004 was probably the best white Christmas in two decades for many western parts of the UK, but even on that occasion snow on the ground was in short supply on Christmas morning itself; approximately 5 million people woke to a snow cover that day, which means that 55 million did not.

For the rest of us, we have to go back to 1981 for the last traditional white Christmas. Much of the UK was snow-covered that day – in some areas it was 30 centimetres (one foot) deep – thanks to repeated heavy falls during the three weeks before the holiday. But 1981 does not appear in the bookmakers' list of white Christmases because the day itself was simply sunny and cold, and no snow actually fell from the sky.

Are you dreaming of a pale-green Christmas?

Another year, having droned on for ages during a radio interview about the difference between the traditional white Christmas and the bookmakers' version of it, I was asked quite an interesting question: 'If snowy Christmases are so rare, then, what is the most common sort of Christmas weather?'

I had a gut feeling about the answer, but I needed to spend a few hours leafing through the record books to make sure. Since the Clean Air Acts of the 1950s and 1960s, widespread fog has been an infrequent visitor over the holiday. Foggy Christmas Days in the last half-century happened only in 1958, 1969, 1973 and 1992. Before that, though, there was a notable run of three very foggy Christmases in 1942, 1943 and 1944, and in London the fog lasted throughout Christmas Eve, Christmas Day and Boxing Day in each of those years.

We do not, of course, have heatwaves in December, but abnormally mild spells have occurred in a number of years, most recently in 1998 when the temperature reached 15°C (59°F) at Torquay, and also 1997 when the same figure was just exceeded at RAF Chivenor in north Devon and at Bristol. The warmest Christmas Day on record was in 1920 when the mercury climbed to 15.6°C (60°F) at Killerton Park on the outskirts of Exeter.

Thunderstorms are very uncommon, having happened over a wide area only in 1947, 1989 and 1990 on the 25th. The 1990 storms were associated with a very vigorous cold front which swept rapidly eastwards across the whole of the British Isles. This front gave most of us a day of really wild, windswept weather with heavy rain and hail. There were also a number of small tornadoes which caused damage in several midland and southern counties of England, especially Somerset and Wiltshire. Damaging gales have also been a feature of several recent Christmases, notably 1997, 1998 and 1999. On Christmas Eve 1997 a ferocious south-westerly gale with gusts reaching 97 knots (111 mph) at Aberdaron on the Lleyn Peninsula tore across Ireland, Wales, northern England and south-west Scotland, bringing down power and telephone lines and leaving thousands of people without power or light over the holiday. Scotland was in the firing line late on Boxing Day 1998, with the Central Lowlands worst hit; Prestwick in Ayrshire recorded a gust of 90 knots (103 mph). And on Christmas Eve 1999 it was Wales, the West Country and the Channel coast which suffered most as winds gusting 85 knots (98 mph) swept high spring tides over coastal defences.

The most common Christmas weather during the twentieth century was blue sky, bright sunshine, and a sharp nip in the air, which happened eighteen times out of a hundred. Sometimes these days occurred in isolation as in 1984 and 1979, but in other years they formed part of a long spell of fine and frosty weather and this was the case several times in the 1960s. So a crisp, crunchy, pale-green Christmas, our parks and gardens sparkling with hoar frost and rime, icy patches on footpaths and side roads but major routes glistening white with the local council's salt, this seems to be the nearest thing we have to normal seasonal weather.

The Great Highland Blizzard of January 1978

The 60-something motorist was dug out of the giant snowdrift on 31 January. It was Tuesday morning; he had become stranded by the blizzard late on Friday night as he travelled between Inverness and Wick, and his car had been buried under six metres (about 20 feet) of snow for more than three days. The Ord of Caithness is a dangerous place in a snowstorm; here the main A9 road between Helmsdale and Berriedale winds through several narrow steep-sided valleys separated by stretches of exposed moorland sloping towards the coastline but 230 metres (750 feet) above it.

Unlike several others who had perished in this ferocious Arctic storm, this man was found alive and kicking because he had known how to survive. He had had some biscuits with him, he had also been able to keep clear an opening in the side of the drift, maintaining a fresh supply of air, and he had the sense to stay inside his car. But – to the delight of the newspapers who told his story afterwards – he was a travelling salesman and he carried a case of ladies' tights and other assorted underwear; he used these samples to pad out his own clothing and to wrap his feet and hands, thus crucially keeping at bay the creeping effect of hypothermia which had claimed the lives of three other motorists.

The winter of 1977–78 does not stand out in the statistical tables as a particularly cold one – indeed, in the last half-century there have been 20 colder winters. However, that particular season occupies a unique place in the meteorological annals thanks to an extraordinary sequence of individual events, including coastal flooding in Lancashire and north Wales in November, a violent gale combined with further coastal floods around 11 January in eastern England which destroyed several Victorian seaside piers (had this North Sea surge coincided with a spring tide it would have created floods worse than the 1953 catastrophe), Glasgow's heaviest snowfall for 31 years on 18/19 January, and during the middle of February one of the West Country's most severe snowstorms of the century. There was also an unexpectedly heavy snowfall over a large part of England on 11 April.

As extraordinary as any of these events was, none of them quite matched the Great Highland Blizzard which raged without a break from the evening of 27 January until the morning of 30 January. A depression raced across the Atlantic between the 25th and 27th, the centre tracking close to the 55th parallel, but as it approached Ireland it decelerated and intensified dramatically. The depression reached its greatest depth on the evening of the 28th with a central pressure of 963 millibars off the Durham coast, and the weather chart on this day was a splendid example of, in weather forecasters' jargon, a 'dartboard', with the bull's-eye over northern England surrounded by a tightly packed mass of near-concentric isobars. As the depression travelled slowly eastwards across the North Sea during the 29th and 30th, in its wake a fierce blast of northerly winds swept the whole country.

The air in the circulation of this depression was not abnormally cold, and around the coastal fringe of Scotland and in the Central Lowlands snow turned readily to sleet and rain so that Scotland's main population centres saw very little accumulation of snow on the ground.

It was, however, a very different story inland, especially north of the main Grampian watershed, where the temperature remained at or below freezing throughout the snowstorm, accompanied by a ferocious north to north-easterly gale. Wet sticky snow clung to power and telephone lines where it froze, soon resulting in failure of the lines and interruption of the services; posts, pylons and trees also snapped under the weight of frozen snow. A dramatic increase in the level of drifting during Saturday the 28th overwhelmed the efforts of the various highways departments to keep even trunk roads open. The snow gradually became lighter and more intermittent during Sunday afternoon and evening, although snow showers continued into Monday 30th.

Because of the severe drifting, it was difficult to assess 'level' snow depths after the storm, but reasonably accurate measurements include 22 centimetres (9 inches) at Lairg in Sutherland, 33 centimetres (13 inches) at Achnashellach in Wester Ross, 58 centimetres (23 inches) at Glenlivet in Banffshire, 60 centimetres (24 inches) at Inverdruie and 65 centimetres (26 inches) at Glenmore Lodge, both near Aviemore. A depth of 90 centimetres (36 inches), including the remnants of earlier falls, was reported from Clashnoir, located roughly ten kilometres (six miles) south of Glenlivet.

Large parts of the highlands were without electricity or telephone contact. Some 200 cars, two buses, and three trains were stranded. Many people had found shelter, and one cottage on the Slochd Pass between Aviemore and Inverness took in 63 travellers. During the four days after the blizzard subsided, dozens of helicopters were mobilized, rescuing hundreds of people from remote sites. The human death toll of six was surprisingly low in the circumstances. However, livestock and wildlife were badly hit, although one sheep was dug out alive as late as 6 March.

Everyone was rescued by 3 February, and the State of Emergency was declared over on this date, but the blizzard had its final say nine months later. According to health authority figures, the birth rate in this period in the Highland Region showed a 30 per cent jump above normal – 190 as against 145.

The Winter of Discontent

Summer weather lingered long into October and November, with misty mornings, warm hazy days and soft breezes. It was the driest autumn in over 250 years of records, with less than half the normal amount of rain over most of the country, and less than a quarter in southern England. The temperature regularly climbed into the 70s in September and October and into the 60s in November, the autumn colours were stunning, and the leaves hung on to the trees long into the dark days of December. It was 1978.

Several of Prime Minister Jim Callaghan's colleagues urged him to go to the country in the autumn of 1978. The gentle, genial – almost complacent – October weather seemed to mirror Callaghan's prime ministerial image, the economic storms of the previous two years had been survived, just, and inflation was in single figures for the first time in several years and was still falling. Political commentators, with hindsight, thought he could have pulled it off and won another term in office.

The opportunity was declined; the weather changed. On 22 November – a date resonant with history in the political world – the large high-pressure system which had graced Michael Fish's weather charts for the best part of three months suddenly shrivelled, like a pricked balloon, and a very cold north wind plunged down from the Arctic to fill the gap. Severe frosts gripped the country during the last few days of the month, snow and hail showers peppered northern and eastern districts, and within 24 hours shirtsleeves gave way to overcoats and scarves.

Thereafter, Callaghan's government was dogged by a sequence of industrial relations problems, strikes were exacerbated by secondary picketing, rubbish piled high in the streets and burials were delayed. Yes, you remember, we called it the 'Winter of Discontent'.

The misery and discomfort were matched by the weather. This was not a truly severe winter in the manner of 1963 or 1947. The country was not snowbound for weeks on end. But we were assailed with repeated surges of Siberian and Arctic air, each of which brought heavy snowfalls, hard frosts, and winds which chilled us to the marrow. Each of the cold spells lasted a week or two, then relaxed its grip, allowing a thaw to get under way, but, just when we thought the worst was over, the temperature dropped again. Only 20 days brought weather from the west or the south-west, from the relatively mild Atlantic Ocean, between the end of November and the beginning of March.

Statistically, this was the coldest winter since the infamous freeze-up of 1962–63, and no winter since has been as cold. The Central England Temperature, a notional figure representing a large chunk of the country stretching from the Thames Valley to Lancashire, for the winter quarter (December, January and February) was 1.6°C (34.9°F), and in the twentieth century only the winters of 1916–17, 1939–40, 1946–47 and 1962–63 were colder. The equivalent figure for 1962–63, the coldest winter since 1739–40, was –0.3°C (31.5°F).

One very sharp but comparatively short cold snap lasted from 28 December to 5 January. The drop in temperature was sudden and followed a period of prolonged rain which resulted in flooding, especially in northern England where York was particularly badly hit. The floodwaters froze, and snow fell on top. A blizzard swept southern England on the 30th, paralysing road, rail and air traffic: Heathrow airport was closed for several days. New Year's Eve was the coldest December day for 40 years, with several places remaining below –4°C (25°F) throughout the day, and 1979 was ushered in with one of the hardest frosts of the 1970s, with minima as low as –16°C (3°F) even in Cornwall. Snow lay for three days in the Isles of Scilly, a very rare event.

The lowest temperature of the entire winter was logged on the morning of 13 January at Carnwath in Lanarkshire, a bone-chilling –24.6°C (–12.3°F) which was just 2.5 degC short of establishing a new UK record. Then on the 23rd another disruptive snowfall hit southern England; 10–15 centimetres (four to six inches) of snow was followed by freezing rain in the London area, and the capital's transport network seized up completely. But perhaps the most severe weather of the winter struck between 14 and 16 February when heavy snow was accompanied by an easterly gale which swept the dry powdery snow into huge drifts. Many towns and villages in East Anglia, the northern Home Counties and the Midlands were cut off for several days, and matters were made worse by widespread failure of the electricity supply.

The old Labour government limped from crisis to crisis through the first three months of 1979, finally losing a vote of confidence on 28 March. Even the five-week-long election campaign which followed was plagued by rain, snow and cold winds, and the only halfway decent weather coincided with a break in campaigning over the Easter holiday. Snow showers fell widely during the opening days of May, including polling day itself – 3 May. The rest, if you'll forgive the cliché, is history.

The wind-chill factor

I have a confession to make. The wind-chill temperature, that ghastly invention which scars some of our radio and television weather forecasts during the winter, was first used on a systematic basis in the UK by . . . me.

When I started forecasting the weather on the wireless in the early 1980s the principal object, agreed upon by the radio station and by me, was to make the forecasts more accessible to listeners by changing the language used, by explaining why the weather did what it did, and by relating it to the everyday life of our listeners. Those, you may remember, were the days when most prognostications indicated that the weather would be 'changeable with showers or longer periods of rain, but there would also be bright intervals', while the phrase 'wet and windy' would never pass the lips of the lads and lasses on the telly. Rather, they would tell us that there would be 'outbreaks of rain, heavy in places, and a fresh to strong wind'.

The introduction of the wind-chill factor was very important in this regard. It had been used regularly in weather forecasts in the USA since at least the 1940s, but no forecaster in Britain had yet made a serious attempt to explain why it felt so much colder at a given temperature when the wind blew strongly.

The concept of wind-chill is a simple one. But measuring wind-chill is a very different kettle of fish, involving what physicists call 'energy flux'. The cold we feel when the wind blows is the result of the loss of heat energy from the human body. When the temperature is low, warm-blooded creatures like us lose heat, but that energy is dissipated more and more rapidly the stronger the wind blows. Strictly speaking, the apparent degree of cold is measured as energy loss per unit area per unit time. The correct scientific units are *joules per square metre per second*, but as the resulting numbers are extremely small, for practical purposes the units used are either *joules per square centimetre per minute or kilojoules per square metre per hour.*

Now, we never hear these words in our weather forecasts, and if we did we would certainly tell the weather presenter to shut up and start speaking English. So meteorologists thought up the simpler idea of *equivalent temperature*. This is the temperature which causes the same energy loss to an unclothed human body when there is no wind blowing. For instance, a temperature of 0°C (32°F) with a steady wind speed of 20 knots (23 mph) results in the same energy loss as a temperature of –16°C (3°F) with no wind at all. This *equivalent temperature* is commonly but wrongly called the 'wind-chill temperature'.

There are, however, many experts who object to the use of wind-chill temperatures in weather forecasts for other than semantic reasons, although they may be perfectly happy with the concept of wind-chill itself. Their objection is that the wind-chill temperature is not a temperature at all, but a scientific abstraction, a notional value, which makes broad assumptions about human physiology. The cooling of the human body also depends on body size and shape, metabolic rate and one's general level of health and fitness, which all vary considerably from individual to individual. Other meteorological factors such as humidity and sunshine also come into play. Furthermore, not many of us are in the habit of prancing about outdoors in winter with no clothes on, even when there is no wind at all.

The objectors also point out that it can be misleading, dangerous even, to talk about wind-chill temperatures of 0°C (32°F) or below in general weather forecasts. Some listeners and viewers might infer a risk of icy roads, damage to plants, and so on, even though the true air temperature is invariably much higher. Further, the use of two 'temperatures' in a weather prediction can only be a source of confusion to the general public. What are we supposed to make of a forecast which tells us the lowest temperature will be –5°C but the wind-chill effect will make it 'feel like' –15°C? We should rightly be scathing about anyone who suggests that a given temperature with a strong wind could 'feel like' another temperature with no wind at all. This confusion is further aggravated by the news media quoting wind-chill temperatures rather than actual ones, because lower readings make better headlines.

Those who take exception to quoting these notional temperatures are quite content to use descriptive terms to emphasize the effect of wind-chill, and it is not beyond the wit of even the more obtuse of our weather forecasters to find an appropriate form of words. For instance: 'the day's maximum temperature will be three degrees centigrade (37 Fahrenheit) but don't forget there is a large additional chill factor thanks to the strong east wind which will be blowing all day . . .' Nor is there any need to resort to that awful expression beloved of some weather presenters: '. . . and the wind will feel cold', implying that it isn't really. I'm sorry, but I don't know anyone who, coming indoors on a perishing day, is likely to say: 'By Jove, that wind feels cold!'

After two years – 1983 and 1984 – using equivalent temperatures in my forecasts, I realized the error of my ways and repented. A year later the Meteorological Office forecasts *started* to use them. I am now more than ever convinced that wind-chill temperatures should not be used in weather forecasts, and in a spirit of meteorological fraternity I cordially invite other weather presenters to join me. Please.

Chapter 7

Twenty-first-century weather headlines

Introduction

Whenever we have a spell of unusual weather these days, we seem to think it is something brand new. There have certainly been plenty of them since that '1' on the calendar flicked over and became a '2'. Many of us know that the third millennium and the twenty-first century did not really begin until 1 January 2001, but the switch from 1999 to 2000 seems to provide our brains with a more natural threshold. It also allows me to include in this chapter the floods (and gales) of autumn 2000 which were the most widespread and destructive to hit England and Wales since the North Sea storm surge of 31 January–1 February 1953, and the great thaw flood of March 1947.

My selection is arbitrary and subjective, but it illustrates the range of weather extremes that the UK experienced in one measly five-year period. Most important, and notwithstanding what some self-styled experts might say about the increasing frequency of extreme events, I have also tried to show that very nearly every one of them was not without precedent.

There is another way to illustrate my point. Let me pick another five year-period entirely at random, 1952–56, the first five complete years of my life, and list the outrageous meteorological extravagances that we had to endure then:

1952

15 Jan: Destructive gale in Scotland, gusts to 108 mph at Stornoway
31 Jan: Level snow 35 cm deep north-east Wales, drifts 1.5 m high
29 Mar: 24-hr blizzard southern England, 25 cm deep in the Chilterns
19 May: Tornado caused much damage at Tibshelf, Derbyshire
1–31 May: Overall, this was the warmest May since 1848
6 Aug: Floods north-west London after deluge, 122 mm rain Borehamwood
15 Aug: Lynmouth disaster, 34 died in flash flood; 229 mm rain Exmoor
1–30 Sep: Coldest September since 1807
27–30 Nov: 30 cm of snow Midlands. Train stuck in 3 m drift Pen-y-Wern
5–9 Dec: Great London Smog, over 7,000 excess deaths in capital
16–18 Dec: Violent gale and heavy snow. Gusts to 110 mph Cranwell

1953

31 Jan: North Sea floods, 307 UK deaths. Gusts to 124 mph Orkney
1–31 Mar: 18 days of fog in Yorks, Lancs and the Black Country
25 May: Widespread thunderstorms, three people killed by lightning
26 Jun: Flash floods Southern Uplands, 80 mm in 30 mins Eskdalemuir
4 Dec: 17°C in London, the capital's warmest December day on record

1954

15 Jan: Severe gale, several deaths, gusts to 97 mph Glasgow airport
29–31 Jan: Snow 35 cm deep Shropshire, Flintshire, Denbighshire
1–7 Feb: Villages in Kent cut off by 2 m snowdrifts; –20°C Welshpool
1 Jun–31 Aug: Worst summer of the twentieth century
20 Sep: Ball lightning observed at Southport
1–31 Oct: Scotland's wettest October since 1903, widespread flooding
6 Oct: Tornado damage at Sprowston, near Norwich
26 Nov: Severe gale, 101 mph gust Stornoway, serious floods in Hull
8 Dec: West London tornado, damage from Chiswick to Highgate

17–18 Dec: Rainfall of 263 mm in 48 hr at Cruadhach, Inverness-shire

1955

1 Jan–11 Mar: Snowy winter, 40–50 days with snow lying in many areas
16 Jan: Complete darkness at 1 pm in London under thick smoke pall
21 Feb: Level snow 90 cm deep north-east Scotland, drifts 9 m high
23 Feb: Coldest night in the UK since 1895, Braemar had –25°C
23–28 Feb: Operation Snowdrop supplied isolated towns by air
23 Mar: Severe gale, much damage, gusts to 94 mph at Scilly
17 May: Exceptionally late snowstorm, 10 cm deep in west Yorkshire
18 Jul: Heaviest daily rainfall on record, 279 mm at Martinstown, Dorset
1–31 Jul: Driest July since 1911; no rain at all in Cornwall, East Anglia
1–31 Jul: Sunniest July on record in the north-west and Midlands
20 Dec: Heavy snow northern England, 30–35 cm snow in Yorkshire

1956

10 Jan: Heavy snow north and east England, 40 cm deep in Lincolnshire
1 Feb: Intense cold, maximum temperature –6°C East Anglia and Kent
10 Feb: Sea frozen along coast of east and south-east Kent
21 Feb: Snow now 35–45 cm deep in east Kent following repeated falls
11 Jun: Highest 2-hr rainfall on record, 155 mm at Hewenden, Yorks
29 Jul: Record-breaking July gale, 93 mph gusts Lizard, much damage
1–31 Jul: London's wettest July in 300 years of records
6 Aug: Piles of hailstones 1 m deep in Tunbridge Wells following storm
1–31 Aug: Wettest August on record in north-west England
1–31 Dec: Dullest December of the century; 1.5hr sunshine at Bolton
25 Dec: A White Christmas, snow up to 22 cm deep Midlands, Essex

There was nothing all that unusual about the 1950s, weather-wise, although it is true that there was a notable concentration of disasters in 1952–53. Nevertheless, any other five-year period during the twentieth century would have produced a list of remarkable meteorological events just as long or even longer. Since 2000, the snows seem to have largely vanished, although the occasional smatterings we get now seem to cause more disruption than did those enormous falls half a century ago; we now have 'killer heatwaves' which were unheard of in the 1950s even though three or four summers were fine and very warm during that decade.

Autumn 2000 – floods in Sussex and Kent . . .

Ever since Noah, flooding has been part and parcel of our natural environment, and that is just as true of Britain as it is of Babylon or Bangladesh.

It is, I am sure, scant comfort to the inhabitants of Uckfield, Lewes, and other towns and villages so badly hit by the October 2000 flood to be told that their disaster was perfectly normal. But it is important to emphasize that there is no need to seek out any 'special' reason for the torrential downpours which caused the rivers to rise so dramatically. Rainfall of this intensity, although rare, is by no means unprecedented; in other words, such a rainstorm falls well within the envelope of meteorological variation which characterizes the climate of south-east England.

Having said all that, the *impact* on our homes and communities of the 2000 flood was undoubtedly much greater than that of a similar flood, say, half a century ago. For this we can blame ourselves: building new estates and roads on flood plains, draining marshes and water meadows, culverting streams, and so on. All these activities contribute to the severity of flooding along a given watercourse, and all these activities have become endemic in recent decades.

Insurance losses were huge. The aggregate of pay-outs approached £200 million. But the recent inflation in weather-related insurance claims is only partly to do with the weather; rather it is a reflection of how well insured we are these days, how efficiently we keep our insurance cover up to date, and how much more costly our property and belongings are to repair or replace. It also tells us a lot about the cavalier way in which planning permission applications which are refused by planning departments are overturned by local councillors, allowing property developers to build new houses, schools, supermarkets and other businesses on the flood plains of rivers. Flood plains are there for a reason – to permit periodic flooding when rivers overtop their banks, and they are flat because of the silt and mud which have been deposited at regular intervals over aeons of time. We all knew this once, and flood plains were never built upon. But since the 1950s, and especially in the last twenty-five years, greed has begotten amnesia.

The deluge of 10–11 October was symptomatic of autumn 2000, not just in Britain but over much of the western half of Europe. A relatively hot August and September (more apparent on the continent than in the UK) had left the waters of adjacent parts of the Atlantic Ocean and Mediterranean Sea several degrees warmer than usual. This unusual warmth injected additional energy into the rain-bearing weather systems heading into Europe from the south-west, making them more intense than normal. From mid-September onwards the weather was very disturbed, with high winds and heavy rain at times. Rivers overtopped their banks in many parts of England and Wales between 14 and 21 September, and Portsmouth was badly flooded when a pumping station broke down on the 15th. Violent cloudbursts hit north-western Italy in late September, and the same storm system that was responsible for 'our' flood caused severe flash floods and mudslides in the Swiss and

Italian Alps from 12 October onwards, resulting in over 40 deaths. Ten days later it was the turn of eastern Spain, with 285 millimetres (11.2 inches) of rain falling in 48 hours at Valencia; the death toll there was 10.

Here in the UK a large number of rivers overflowed during this lengthy period, but nowhere else in the country was remotely as badly hit as Sussex and Kent and adjacent parts of Surrey and Essex. This was the result of an exceptional coincidence of meteorological factors. A line of thunderstorms formed along the boundary between two contrasting airmasses stretching approximately from Brighton to Maidstone to Southend. This string of storms became aligned along the flow of the wind (from the south-south-west) for over 36 hours so there was no impetus for it to move away. Furthermore the warmth of the English Channel provided sufficient additional energy to the system to cause new storms to generate as the older ones gradually weakened. In effect the downpour was self-perpetuating. Consequently, the area which received abnormally large quantities of rain was quite restricted, with 100 millimetres (four inches) or more falling on the tract of countryside extending from the north-eastern outskirts of Brighton to Ashdown Forest between Crowborough and East Grinstead. At Plumpton, near Lewes, 174 millimetres (6.85 inches) of rain fell in 48 hours, a total which occurs somewhere in England once every five years on average, although at a specific location it may not occur again for many centuries. Small as the deluged area was, the floodwaters fed into catchments of several rivers, including the Adur, the Rother, the Eden, the Medway and the Ouse, but the Ouse and its tributaries collected by far the most, and that is why Uckfield and Lewes suffered so much.

How does this quantity of rain compare with Britain's biggest-ever downpours? The greatest fall ever recorded in the UK was at Martinstown, near Dorchester, when 279 millimetres (11 inches) of rain fell on 18 July 1955. Somerset was twice hit: on 28 June 1917 some 243 millimetres (9.5 inches) was deposited on Bruton in 18 hours, and on 18 August 1924 Cannington, near Bridgwater, collected 234 millimetres (9.4 inches), most of which fell in five hours. The famous Hampstead storm of 15 August 1975 produced 171 millimetres (6.7 inches), almost the same as Plumpton's two-day total, in just 150 minutes, and the Boscastle storm of 16 August 2004 dumped 200 millimetres (7.9 inches) on the moors above the village in four hours.

None of this is meant to diminish the severity of the October 2000 flood which was reckoned by the experts to be the worst to hit the south-eastern corner of England since the disaster of September 1968. On that occasion, a much larger region was hit, and slightly more than 200 millimetres (eight inches) of rain fell inside two days at Tilbury and Stifford in south Essex. The *Daily Telegraph*'s main headline was 'Floods Devastate South – Hundreds of Homes Evacuated' but this played down the real situation, for, the day after the rains stopped, the floodwaters covered over 250 square kilometres (100 square miles) of land and had penetrated 25,000 homes. For the best part of a day, London was completely inaccessible by road from the south and south-west. The River Arun in West Sussex rose from a trickle to a five-metre (17-foot) deep torrent, while the River Mole in Surrey was carrying more than four times the previous record quantity of water.

The 1968 flood eclipsed that of October 1960 in south-east England, but the 1960 event was much worse further west – in the West Country and south Wales. Nevertheless, there was extensive flooding in Hampshire, west Surrey and west Sussex around 9–10 October that year. A little more recently, an extraordinary deluge struck east Kent on 20–21 September 1973, with 191 millimetres (7.52 inches) of rain collected in 18 hours at Stourmouth, near Sandwich, and more than 100 millimetres (four inches) fell over a large area extending from Folkestone and Ashford to the Isle of Thanet. One person died in the resulting floods. In much more recent times, repeated flooding hit Sussex and Hampshire during December 1993 and January 1994, Chichester being particularly badly affected, while August 1996 brought a severe but localized flood to the Folkestone–Ashford district.

In the wake of the October 2000 downpour, experts warned that the fact that the Uckfield/Lewes flood occurred at the beginning of the season meant that any further heavy rain during the winter half-year would bring renewed flooding. They were proved right.

. . . and in southern Europe

The flash floods and mudslides in the Swiss and Italian Alps in mid-October 2000, just like the flooding in south-east England a few days earlier, yet again saw many news bulletins make the link between extreme weather events and so-called global warming. The link is made without equivocation or explanation, yet no reputable scientist would make a similar association without a bucket-load of reservations.

Let me reiterate that, although the warming trend in our planet's climate may well bring about more weather disasters, it is completely wrong to blame individual events such as the recent floods on the changing global climate. Unless the meteorological measurements exceed previous extremes, or the frequency of their occurrence increases beyond previously known limits, there is no need to seek any special explanation for these events. They are simply part and parcel of normal climatic variability, and this is entirely typical of all kinds of climate throughout the world.

Let us look more closely at the rainstorms which hit the Alpine region during that month. The bald statistics are impressive. At Locarno, at the Swiss end of Lake Maggiore,

308 millimetres (12.13 inches) of rain fell in the week up to 16 October, which is nearly twice the local average for the whole of October. It is also the equivalent of six months' rain in London, a statistic which our tabloid news outlets may find impressive but which is otherwise supremely irrelevant. At Nice, on the French Riviera, 209 millimetres (8.21 inches) of rain fell in six days; in Italy 172 millimetres (6.75 inches) fell in five days at Turin.

The climate of southern Europe at this time of the year shows very wide variations. Some Octobers are completely rainless while others are characterized by repeated torrential downpours. Three or four times the monthly normal is by no means rare there. Scarcely an autumn passes without a destructive flood somewhere in the region, whether it be in the Alps, in southern France, in Spain or in Italy. The October 2000 flood was the worst in northern Italy only since November 1995, when 70 people died. In October 1973, 350 died in devastating floods in eastern Spain, and in Barcelona 475 drowned 11 years earlier. November 1966 brought Italy's worst weather disaster of the century, with 113 dead and irreparable damage to historic buildings and art treasures in Florence, while in 1963 torrential rains triggered a landslide which in turn caused a dam to fail at Longarone in the north-eastern corner of Italy, killing almost 2,000. More recently, deadly flash floods ravaged valleys leading from the southern flank of the Massif Central in France in September 2002 and again in early December 2003.

When Atlantic depressions penetrate the countries of south-western Europe during early autumn they draw into their circulation very warm and very humid air which may have stagnated for several days over the Western Mediterranean Basin. The warmth and moisture re-energize the depression which now turns into a very efficient rain-making machine, and the steep and broken topography of the region intensifies the rain-making process. In these countries, just as in the UK, human activity – in particular the untrammelled growth of towns and cities on flood plains – puts many thousands of people's lives at risk and considerably exacerbates the level of property damage.

Severe gale and more floods: 30 October 2000

John Prescott informed the House of Commons on 31 October 2000 that the previous day's gale was 'according to the Met Office, the worst storm since 1987'. Mr Prescott's speaking style may lead us occasionally to wonder out loud who his scriptwriter is. However, these statements to the House are normally written by civil servants, and the extraordinary error quoted above certainly leads me to wonder whose fault this particular script was.

That the news media tried to compare the October 2000 gale with the Great Storm of October 1987 does not surprise me. Leaving aside for the moment the dozen or so gales of similar or greater severity which affected England and Wales during the preceding ten years, our news outlets have largely ignored the violent gale of 25 January 1990. It is the great forgotten storm of recent decades.

Curiously dubbed by meteorologists the 'Burns' Day Storm' – my Scottish friends assure me that there is no such thing as Burns' Day, only Burns' Night – this was in many ways a worse national disaster than the 1987 storm, and it was a couple of orders of magnitude more severe than the October 2000 one. All parts of England and Wales together with Northern Ireland and southern Scotland were swept by the high winds during the 1990 gale, in contrast to the 1987 storm which was confined to the south-eastern quarter of England. The death toll of 47 compares with 18 in October 1987 and 8 in October 2000. Insurance losses, approaching £4 billion adjusted to current prices, were twice those of the 1987 storm. The insurance industry, not normally known for underplaying a weather disaster, believe that their liability from the 30 October 2000 gale (not including the floods) may end up close to £100 million: I heard one of their representatives describe it as 'a non-event' without so much as a flicker of embarrassment.

The surprisingly small amount of structural damage on 30 October 2000 can be understood when one considers that the force exerted by a strong wind increases geometrically as the actual wind speed rises arithmetically. For example a change in wind speed from 60 to 70 knots is an increase of 16 per cent, but the force of the wind increases by 30 per cent.

As for those destructive gales which hit England and Wales between 1990 and 2000, it is amazing how quickly we forget such extreme events unless we are directly affected. For instance, that of Christmas Eve 1997, completely forgotten by most of us, will live long in the memory of tens of thousands of people in Wales and northern England who survived a joyless Christmas with neither heat nor light.

The wettest autumn on record

William of Orange was still on the throne, England and her allies finally emerged victorious against France in the War of the Grand Alliance, the fairy stories *Sleeping Beauty* and *Little Red Riding-Hood* were first published in Paris . . . and an Essex clergyman called William Derham installed a rain-gauge in his Upminster garden. The year was 1697, and Derham's assiduous daily measurements from 1697 until 1716 mark the beginning of an unbroken record of rainfall in the London area which now extends beyond three centuries. Gradually, other observers established gauges around the country, and from 1727 onwards there were enough of them to calculate an approximate average representing the whole of England and Wales.

It is thanks to these pioneers in meteorological observation that we can now say with some certainty that autumn 2000 was the wettest on record – the wettest for at least 300 years. A downpour on 23 November finally took us beyond the previous record established way back in 1852, with a full week of the statistical season still to go. Then, as in 2000, all three months of the autumn quarter were very wet, but the wettest of all in 1852 was November, especially in the southern half of England. The River Thames inundated extensive areas of Oxfordshire, Berkshire, Buckinghamshire, Middlesex and Surrey for a period of two to three weeks; this disastrous event came to be known as 'The Duke of Wellington's Flood' because the heaviest of the rains began on 14 September, the very day of the Iron Duke's funeral.

In my *Sunday Telegraph* commentaries, I have often written that, however unusual or extreme individual weather events and seasons may appear, we do not have to scrabble around looking for special explanations if they fall within the range of previous experience. None of the gales and downpours that we endured during October and November 2000 were unprecedented, so there was no need to seek an explanation outside the normal variability of the British climate. However, in aggregate, we now see that the season's rainfall had no precedent during the three-century-long instrumental record of our country. It is reasonable to say now that Britain's exceptionally wet autumn of 2000 was probably no more than an extreme example of our climatic variability, but it is possible that there were additional contributory factors. It would be unwise to express any certainty about these notional additional causes until long-standing records are broken widely, frequently, repeatedly and by increasingly wide margins.

If pressed, a meteorologist would point to the seemingly endless succession of Atlantic depressions which travelled across the British Isles from mid-September onwards, and to the remarkably persistent 'blocking' high-pressure system which was established over European Russia during the same period. This 'blocking high' caused the Atlantic lows to decelerate as they moved into Western Europe which in turn meant that the rain-bearing fronts lingered longer over the UK than is normally the case. Warm ocean waters in the Channel and South-west Approaches also provided additional energy to feed these Atlantic disturbances, thus swelling the rainfall totals in southern England, especially Sussex and Kent.

The heavy rain which returned during the closing days of November ensured that it was the wettest autumn by a wide margin. The final figure, averaged over England and Wales, was 502.7 millimetres (19.79 inches), 93 per cent above the long-term mean, and wetter than the previous record-holder, 1852, by 47 millimetres (almost two inches). At Plumpton, East Sussex, the season's total of 735 millimetres (almost 29 inches) was the equivalent of nine to ten months' rain in an average year. Scotland and Northern Ireland also had a rain-soaked autumn, although here it was not a record-breaker; it was the wettest in Northern Ireland since 1954, in Scotland since 1984.

Here are the ten wettest autumns (1 September–30 November) of the last 300 years, averaged across the whole of England and Wales:

Rank	*Year*	*Millimetres*	*Inches*	*% of normal*
1	2000	502.7	19.79	193
2	1852	455.8	17.94	175
3	1729	440.0	17.32	169
4	1960	438.6	17.27	169
5	1935	424.0	16.69	163
6	1770	402.4	15.84	155
7	1772	400.6	15.77	154
8	1875	399.1	15.71	153
9	1768	398.7	15.70	153
10	1976	397.0	15.63	153

The long-term mean rainfall for the autumn quarter is 260 millimetres (10.24 inches).

February and March 2001: the rains return

After a three-week rest, the rain came back again during February. A prolonged downpour hit southern England and East Anglia on the night of Wednesday 7 and Thursday 8 February 2001, depositing over 25 millimetres (an inch) of rain over a wide area.

As if the gods were targeting those districts worst hit by the autumn and early-winter floods, the heaviest rain homed in on East Sussex, Kent and south Essex, leading to renewed flooding on dozens of rivers in the region. Some 50 millimetres (two inches) of rain fell across the eastern half of Kent, and 56 millimetres (2.21 inches) was recorded at Folkestone between midday on the 7th and 6 pm on the 8th. That is the equivalent of one month's rainfall in an average year.

In the first ten days of February approximately 90 millimetres (3.5 inches) of rain was recorded in Kent and East Sussex, adding to the unprecedented quantities of water which had fallen here since the middle of September. In fact, at Herstmonceux, near Hailsham, the total rainfall for the five-month period from 11 September to 10 February was 1,134 millimetres (44.65 inches) compared with the long-term average of 362 millimetres (14.25 inches) – in other words more than three times the normal amount. Astonishingly, it is more than the September–February average rainfall in Fort William, the UK's wettest town. And lest any hardy highlander deride soft southerners for not being able to cope with such piffling amounts of water, remember that while highland river catchments are designed to cope with such large quantities of rain, the much gentler contours and sluggish streams of south-east England are not.

The landscape of the Home Counties, especially of the chalk hills such as the North and South Downs and the Chilterns, changed too. Extensive tracts of low-lying land were partially flooded from October to April, and even the boulder clay, a glacial deposit, of the high Chilterns was thoroughly waterlogged. Countless chalk springs, dry since time immemorial, sprang into life, feeding new streams in valley bottoms which had been untouched by running water for many decades. One such crossed a main road in Surrey, another ran through a public park in Sussex, yet another travelled through the car park of a Buckinghamshire public house. The rebirth of these springs was not entirely due to the excessive rains: the water table in the chalk has been rising in recent decades thanks to reduced abstraction for domestic and industrial consumption. Abstraction reached a peak in the 1960s and 70s, causing many streams to dry up completely, but new reservoirs distant from the Home Counties now provide much of the water for these areas.

A record number of records?

Waterlogged Britain squelched through the first three months of the year 2001 with precious little relief from the downpours. True, there was a two-week-long respite in the middle of January when frost and snow temporarily replaced the rain, but the massed ranks of Atlantic depressions on our newspaper weather charts soon crowded out the timid anticyclone, and we had no further settled spell until early April.

In the 12 months to the end of March 2001, we had, averaged over England and Wales, the wettest April since 1756, the wettest May since 1983, the wettest September since 1981, the wettest October since 1903, the wettest November since 1970, and the wettest March since 1988. More rain fell during the autumn quarter of 2000 than in any similar period since national rainfall records began in 1727, the calendar year 2000 was the wettest since 1872, the total rainfall during the autumn–winter half-year (that is, September 2000 to February 2001 inclusive) was unprecedented, as was the 12-month total from April 2000 to March 2001. For all I know, the number of new records established was a record.

The figures for that 12-month period sum things up nicely. Taking a geographical average over England and Wales, the aggregate rainfall was 1,340 millimetres (52.76 inches) compared with a long-term normal of 919 millimetres (36.20 inches). Put another way, 46 per cent more rain fell during that period compared with the average for the standard reference period. A statistical analysis of all available records indicates that such a large excess can be expected to occur once every 500–750 years on average, assuming that the rainfall climate of the UK is static. There is no statistical evidence, not yet anyway, to support the notion that Britain is becoming wetter or drier, nor, contrary to popular belief, is there yet any indication that the year-on-year variability in rainfall is changing or has changed significantly. The previous record for a 12-month period was during the calendar year 1872 when the total rainfall was 1,285 millimetres (50.59 inches).

But these national figures tell only part of the story, because the greatest excesses in rainfall since April last year have repeatedly occurred in southern counties of England, while some other parts of the UK have actually recorded a deficit. In London, a total of 995 millimetres (39.19 inches) of rain was collected during the 12 months beginning 1 April 2000 compared with the normal of 610 millimetres (24.02 inches), an excess of 63 per cent. Rainfall was first systematically recorded in the capital in 1697 and this figure has never been approached before. At Herstmonceux, East Sussex, 1,496 millimetres (58.91 inches) fell during the same period, more than twice the normal amount. The statistical technique used to calculate the return period of such an aggregate goes off the scale.

Meanwhile, everyone seemed to have forgotten northern and western Scotland, which enjoyed one of its best summers in recent decades in 2000. The rainfall total from April 2000 to March 2001 at Wick, Caithness, stood at 712 millimetres (28.03 inches), some 10 per cent *below* the normal there.

It's a long story

The River Thames grew longer in 2000 and 2001, and that is official. This was not because someone had been busy with a tape measure, or because of some bizarre new Euro-regulation, it is simply a consequence of the unprecedentedly wet weather which southern England experienced during 2000 and 2001.

Over the Thames Basin there was nearly twice as much rain as normal between September 2000 and March 2001. Whatever the amount of rain that falls in a given year, some of it feeds directly into the rivers, but much percolates into the water-bearing rocks – aquifers – beneath the surface of the ground. These aquifers, chiefly comprising chalk, limestone and sandstone, sit upon clay strata which are effectively impervious to water, so the rainwater collects in the more permeable rocks above. The level to which these aquifers become saturated is known as the 'water table'; it rises and falls a little with the seasons, and it is highest of all at the end of a wet winter.

The water table does not go on rising for ever, because the aquifers are not perfectly watertight; they leak. Water will always try to find its own level, and in the irregular geography of chalk, limestone and sandstone hills the water table will occasionally meet the outside world. Where this happens the water seeps out: we call it a spring.

Springs feed streams; each tributary of the Thames carries water from a series of springs, and the headwaters of the Thames itself are no different. The source of the Thames is traditionally regarded as being in Trewsbury Mead, a field adjacent to both Trewsbury House and the A433 Fosse Way, some seven kilometres (four miles) south-west of Cirencester and two kilometres (a mile and a bit) north of Kemble. A stone marker was installed by the Conservators of the Thames in 1974, and a circle of smaller stones identifies the spring itself.

In modern times this spring is normally dry, and local people say that it flows for a few months at the end of our wetter winters – say once every three or four years.

In a letter to one of the Royal Meteorological Society's journals, society member John Miller described a visit on 10 November 2000, quite early in the wet spell, when he found the spring bubbling up into a sizeable pool. He also found several other springs upstream of the official source. During subsequent weeks the water table rose further and it is believed that springs appeared over a kilometre to the north-west.

It was not until late 2003 that the Thames began to shrink significantly again as the water table dropped sharply after an extended nine-month long dry period culminated in an exceptional drought during August, September and October of that year.

Snow in January? Whatever next?

During the second week of January 2003, the news media excelled themselves with extensive coverage of the weather on radio and television, while there were pictures and comment on several newspaper front pages. And all because five centimetres (two inches) of snow fell in central London.

There was nary a word about the 20-centimetre (eight-inch) snowfall in parts of Northumberland and County Durham, or the 25 centimetres (ten inches) that fell over the hill country behind Wrexham in north-east Wales, no reports of the severe gale and rough seas which pounded the south coast of Cornwall, and precious little about the very low temperatures recorded in the Scottish highlands.

Even the snowfall in the Home Counties on the morning of Wednesday 8 January affected a relatively small area. A narrow band of snow developed over the Thames Estuary around dawn, and travelled south-eastwards across south

Essex, Greater London and north Surrey before petering out over Hampshire. A second band took a parallel path across north and west Kent, east Surrey and parts of Sussex around mid-morning. In these areas the snow was typically five to ten centimetres (two to four inches) deep, but a small area of the North Downs between Gravesend, Bromley and Sevenoaks, together with the Thurrock–Rainham district of south Essex, caught up to 15 centimetres (six inches). Elsewhere in the south-east there was just a dusting.

Reports that this was the heaviest snowfall in the region for 12 years were misleading. Although it was probably correct for the very heart of London, just a few kilometres away in Hampstead and Highgate there had been heavier falls on 2 December 1997, 15 February 1994, and 7 January 1994. In the northern outskirts there had been more snow as recently as late December 2000.

We have grown to expect widespread dislocation of road traffic and disruption of rail services when a little snow falls on our conurbations, and we grumble at the authorities whom we blame for failing to keep things moving. 'It doesn't happen in other countries,' we say. That may be true of cities like Montreal or Stockholm or Zurich where winter snows are commonplace. But the chaos of a rare snowfall happens just as much in less snowy cities such as Paris or Washington as it does in London. A month earlier, two centimetres of snow fell in Tokyo – rare in December – resulting in severe delays to commuter trains, the cancellation of 49 domestic flights, and 157 people taken to hospital with broken bones, bruised limbs or heart attacks brought on by the exertion of clearing the snow.

Temperature records tumble: April 2003

The mid-April heatwave of 2003 was a true record-breaker. Too often we hear the weather people on the television telling us that such-and-such a record has been broken, only to discover by examining the small print on the screen that 'records began in 1989' at that particular location. Sometimes even that qualification is not there, although it ought to be; 'too much information' is the usual excuse for omitting it.

Not so on Wednesday 16 and Thursday 17 April. Although no new UK record was established, a large part of the country enjoyed a degree of warmth not experienced in April since systematic temperature recording began a century and a half ago. The sky was practically cloudless, the haze which had prevailed early in the week seeped away steadily, and the sunshine grew stronger and stronger. A soft breeze blew from the south or south-east, drawing up air which had originated over north-west Africa.

The really warm weather began on Monday when the mercury soared into the 20s C (70s F) locally in western Scotland. The high temperatures became much more extensive on Tuesday when 24°C (75°F) was recorded at places as far apart as London, Cardiff and the north-west highlands of Scotland. For much of England and Wales the hottest day of the spell was Wednesday. The maximum reading of 27.4°C (81.3°F) obtained at Stratfield Mortimer, located between Reading and Basingstoke, was the highest in April anywhere in the UK since 1949. Indeed, a higher temperature has only been recorded on 16 April 1949 when 29.4°C (85.0°F) was registered at Camden Square in London, and in April 1893 when Stratfield Turgis, also near Basingstoke, logged 29.0°C (84.2°F) on the 18th.

By Thursday the hottest weather had transferred to western and northern Britain, thanks to a freshening easterly breeze. Lochcarron in Wester Ross recorded 26.9°C (80.4°F), a new Scottish record; Prestatyn reached 27°C (80.6°F), the highest in Wales since 1893; Portrush logged 22.8°C (73.0°F), the highest in Northern Ireland since 1984; and Belmullet touched 24.4°C (75.9°F), the highest in the Irish Republic also since 1984.

There were some truly exceptional readings on some of Scotland's outer islands, for instance, 22.3°C (72.1°F) on South Uist in the Western Isles, 22.1°C (71.8°F) at Rackwick in Orkney, and 19.3°C (66.7°F) at Sella Ness in Shetland; many years pass here without reaching these levels even in high summer. North Sea coasts missed out on the heatwave: throughout the week onshore winds held temperatures close to 10–12°C (the low 50s F) all the way from Lowestoft to Peterhead.

Did anyone at the time suggest that this early surge of heat foreshadowed the record-breaking August heatwave to come? If they did I didn't notice. It has to be said, though, that many of our hottest summers – in the last 30 years at least – were preceded by one or more spring hot spells. This was certainly true in 1976, 1984, 1989, 1990 (repeatedly so), 1995, and to a limited extent in 1997. It was untrue only in 1975 and 1983.

'Killer heatwave'

The headline was, for once, not sensationalist. The withering heat of August 2003 across practically all of western and central Europe seemed to go on for ever, the stifling nights were, if anything, more difficult to bear than the sweltering days, and people dropped in their thousands. France suffered most. Over a large part of that country the temperature climbed into the 40s C (above 104°F) on 11 successive days and remained above 22°C (72°F) on 11 successive nights. The nights were worst in urban areas where the concrete, stone and asphalt re-radiated the heat of the day after dark, and in Paris for instance the mercury remained above 27°C (81°F) all night on several occasions. This was a natural disaster without parallel in Europe in modern times: in France alone almost 15,000 died, across the western half of the continent the death toll was nearer 25,000, and most of them were elderly people unaware that they were becoming progressively dehydrated by the relentlessness of *la canicule*. For those of us who experienced it and lived to tell the tale – yes, I was there on holiday – it was a torture that none would want to go through again.

Notwithstanding over 2,000 deaths and a new national temperature record, Britain lay on the periphery of this catastrophic heatwave and many people actually enjoyed it. Although the hot spell in the UK lasted from 2 to 18 August, the temperature reaching at least 27°C (into the 80s F) somewhere in the country on each of those days, it began slowly and ended gently, and there was a hiccup in the middle. A temperature of 30°C (86°F) or more was obtained on ten days in a row from the 3rd to the 12th, but after a day of great heat on Wednesday 6th, with 36.1°C (97.0°F) logged at Kew Gardens, the next two days were markedly less hot, although after that the mercury again soared, eventually reaching uncharted territory.

The record was broken on Sunday 10th. Highest readings that afternoon were 38.1°C (100.6°F) at Kew Gardens, 37.9°C (100.2°F) at Heathrow Airport and at Aldenham School, near Watford, and 37.8°C (100.0°F) at Wisley Gardens in Surrey. These figures were all scrutinized and accepted in a paper in *Weather* magazine in 2004; a higher reading made at Faversham – the Brogdale Horticultural Trust – in Kent was rejected because of the excessively enclosed character of the recording site.

By the 10th the hottest weather was confined to the south-eastern corner of England – all the observations of 30°C or more were found to the right of a line drawn from Exeter to Lincoln. The previous day was the hottest over most other parts of England, Wales and southern and eastern Scotland, and it was on this date that the temperature climbed to 32.9°C

(91.2°F) at Greycrook, which is located in the Tweed Valley between Kelso and Galashiels, in the Scottish Borders. This established a new Scottish record. No new records were set in Wales or Northern Ireland, nor indeed in the Irish Republic.

The heat seeped slowly away from southern England during the 11th, 12th and 13th, although the temperature still reached 29°C (84°F) in the London area on the last of those dates. But the drought intensified over the following two and a half months.

Coping with the drought

Dribs and drabs of rain affected many parts of the country during the first week of September 2003, but the quantity of water was insignificant in most parts of England and Wales; indeed, the long dry spell was to continue for another six weeks. At several places in East Anglia and the East Midlands there had been barely two millimetres (less than one-tenth of an inch) of rain between 31 July and 20 September, and that is less than 3 per cent of the normal amount. There have been similar or even longer dry periods in the past: in 1959 for instance there was no rain for 59 consecutive days at Finningley, near Doncaster, between mid-August and mid-October.

One of the most surprising things about 2003's hot, dry summer was that the water companies were almost completely silent. There were very few hosepipe bans in operation, and no one spoke about water shortages. A few commentators mistook this quiet confidence for reckless complacency.

The previous long, hot summer, 1995, will be remembered by many as the one when Yorkshire Water got into terrible trouble, having to ferry water in tankers to several towns and cities where many thousands of domestic users had their supply cut off. Spokesmen for the company itself, for the water companies' association, and even John Gummer who was then Secretary of State for the Environment, blamed 'the unprecedented drought'; my colleagues at BBC radio had to physically restrain me from marching into a studio to accuse them of being thoroughly economical with the truth.

It is true that there was a marked shortage of rain between April and August in 1995, but it is also true that the previous winter had been very wet, and Yorkshire's reservoirs were brim-full at the beginning of April that year. How the then management of Yorkshire Water managed to lose so much water in five months is beyond me, and the claim that the shortages were due to a lack of rainfall are given the lie by the experience of 2003.

Slightly more rain fell between April and August 2003 compared with 1995, but in 2003 the drought had begun two and a half months earlier, during the second half of January. Averaged over England and Wales, the February–August rainfall was 331 millimetres (13.03 inches) or 70 per cent of normal in 1995 compared with 368 millimetres (14.48 inches) or 77 per cent in 2003. The January–August figures reversed the order: 495 millimetres (19.49 inches) or 88 per cent in 1995 and 453 millimetres (17.83 inches) or 79 per cent in 2003.

We should, I suppose, have been grateful that, quietly and efficiently, the water companies – Yorkshire Water in particular – at last appeared to have got their act together.

An inch of snow brings chaos

'GRIT-GATE', the *Birmingham Evening Mail* called it. Yes, Birmingham, along with several other towns and cities enjoyed a brief taste of winter at the end of January 2004 as snow fell heavily for an hour or so. Alongside the 'Big Freeze' headline was a picture of a main road with lines of stationary vehicles and very nearly two centimetres (almost an inch) of snow on the ground. Wow!

The cold snap had been forecast a week in advance, and the snowfall on Wednesday 28 January had been adequately predicted the day before in almost every respect – the time of its arrival, the areas which would be affected, the probable quantity, and the fact that it would be followed by a hard overnight frost. Snow had fallen in many parts of the country 24 hours earlier, with accumulations of around 15 centimetres (six inches) in Lincolnshire and north Cambridgeshire; the Wednesday snowstorm followed a brief thaw with scattered rain which diluted the salt which local authorities had already spread on most roads.

While it was snowing during that brief cold snap, people (both journalist colleagues and real people) kept asking me what on earth was happening to the British climate. Rather than bite my lip in exasperation, which is what I used to do when confronted with such dumb questions, I have now honed my response to perfection. 'It's called January,' I say, with exaggerated irony.

It does worry me occasionally that our responses to normal winter weather are now so intolerant, and seem to be growing more hysterical by the year. I wonder what will happen when we next have a major snowstorm over lowland Britain. Global warming or no, we will get one sooner or later. The last time it happened, in February 1991, the media launched a frenzied attack on British Rail after some poor soul uttered the immortal line, 'It was the wrong kind of snow.' The preceding December, when the Midlands motorway network seized up for almost 24 hours, Home Secretary Douglas Hurd instigated an enquiry to determine whether local authorities had performed their road-clearing duties properly.

After the extraordinary gridlock in London and the south-east after a fairly modest snowfall in January 2003 when some motorists were stranded in their cars for 18–24 hours on the M11 motorway, highways authorities reviewed their procedures for dealing with snowy weather, and foolishly claimed it would never happen again. The simple fact is that there is a certain sequence of weather – rain, then snow, then freeze – which no gritting pattern can hope to neutralize, especially if it happens to coincide with the morning or evening rush hour.

These days urban Britons are insulated from the weather for most of the time and seem to regard it as an infringement of their human rights when severe weather disrupts their routine. Instead of whingeing, perhaps we should, like our forebears, just shut up and get on with it.

The Boscastle flash flood

Summer 2004's frequent heavy downpours have emphasized how powerful and destructive rainstorms can be, even in our equable climate. Occasionally, though, storms of truly tropical intensity strike, and the havoc which follows is all the more distressing and bewildering because of their rarity. The cloudburst over Cornwall on 16 August 2004 was one of these.

Contrary to media reports, serious flooding has occurred before at Boscastle: in late October 1996 when ex-hurricane 'Lili' delivered a combination of heavy rain and high tides which inundated the lower part of the village; in the Junes of both 1958 and 1957 when sudden floods developed in circumstances similar to 16 August 2004; and in July 1847 when a much broader region was affected.

The short, steep valleys of north Cornwall and north Devon are particularly vulnerable to localized summer downpours. They collect water efficiently from the surrounding moors (more efficiently now that many of the slopes have been cleared of woodland and scrub), channel it rapidly into the main stream, and take it all out to sea in a matter of three or four hours. Because of their almost instantaneous response to a sudden cloudburst, these valleys are known in the trade as 'flashy catchments' and they produce true 'flash floods'. Now we all know what a real flash flood looks like, perhaps our traffic reports and news bulletins will stop using the term to describe any large puddle which collects in a dip in the road after a heavy shower.

In the rush to find a scapegoat, climate change being the favoured one, most commentators ignored the fact that we were long overdue one of these catastrophes. There was a serious flash flood in mid-Wales two years ago, but before that we have to go back to Helston in June 1993 and Halifax in May 1989. The 1950s and 1960s gave us several major West Country floods, including the Wadebridge and Camelford flood in June 1950, the Lynmouth disaster in August 1952, another Camelford flood in June 1957, Porlock in July 1959, Wadebridge again in June 1963, and large parts of Somerset and east Devon in July 1968.

It took several days before the quantity of rain responsible for the Boscastle disaster was accurately ascertained. We now know that the focus of the storm lay over the moors five or six kilometres (about three miles) east of Boscastle itself; 200.4 millimetres (7.89 inches) fell at Otterham, practically all of it inside four hours. At the peak of the downpour, 90 millimetres (just over 3.5 inches) fell in one hour. Both of those figures represent very rare events for any location in the UK, but neither quite approaches the existing long-standing British record. However, the most extraordinary statistic of all is that no human life was lost.

Sea defences breached in Cornwall and Devon

The gales and high tides which caused so much damage in the West Country on Wednesday 27 and Thursday 28 October 2004 were associated with an exceptionally deep depression which took up residence in the South-west Approaches for several days.

The depression had developed the previous weekend near

Bermuda, and travelled across the Atlantic Ocean in unusually low latitudes – between 35 and 40 degN – before turning north-east and then north towards Cornwall. It was during this last part of the depression's journey that it deepened explosively, its central pressure dropping to 950 millibars (27.99 inches of mercury) on Wednesday afternoon, 27 October. Depressions of this depth are rare enough in the northern Atlantic in October, but this was only the second occasion in the last 30 years that the pressure had dropped so low at this time of the year south of the 50th parallel.

The depression's circulation occupied a huge area, and the resulting rough seas caused damage not only in south-west England and south Wales, but also in eastern and southern Ireland, in western France and northern Spain, and along the Portuguese Atlantic coast. Unusually heavy swells were even noted in Morocco and the Canary Islands, providing superb breaking waves for surfers, but dismaying holiday-makers who merely wanted to lie on the beach in the sun.

The strongest winds, averaging 60 knots (70 mph) with gusts to 80 knots (92 mph), were reported by ships in sea area FitzRoy off north-west Spain. Along the coast of south-west England sustained winds of 40–45 knots (46–52 mph), gusting to 60 knots (70 mph), were observed; such a gale is certainly unpleasant but winds of this strength are typical of an average autumn or winter gale. The extensive damage which occurred to sea walls and harbours in south Cornwall and south Devon, described by locals as the worst for 40 years, therefore requires some explanation.

Several factors came into play. Most important was the coincidence of strong winds and the spring tide. The strong winds blew from the south-east, an unusual direction for a severe gale in British waters, and this exposed the Channel coasts of Cornwall and Devon – generally less well protected than the Atlantic coast – to the full force of the waves. The very high natural tide was augmented by a wind-driven tidal surge which piled water into the bays and inlets along the coast. In addition to this, the exceptionally low atmospheric pressure – less weight of air pushing down on the ocean surface – meant that water levels were already higher than average. Finally, the huge waves driven by the wind were themselves augmented by a massive swell running up from the south-west which was caused by the high winds off the Spanish coast. Thanks to the high water levels these waves were breaking much closer to the sea defences than they would have done on an ordinary tide, and therefore subjected those defences to a much more severe pounding than might have been expected.

The authorities confirmed what most local people already knew: this was the most damaging gale along the south Cornwall coast since April 1962. The cost of repairing and replacing damaged sea walls and other defences, rebuilding washed-out roads and harbours, together with insurance payouts to flooded householders and to fishermen and recreational sailors who had lost their boats, was estimated to be over £50 million.

Chapter 8

Climate-change jottings

Introduction

My first scribblings about the greenhouse effect date back to 1986, about two years before the world's media cottoned on to the fact that here was a 'sexy' news story that would fuel the embryonic millennial angst and thus provide material for the rest of the twentieth century at least. And with the advent of colour in our newspapers coinciding with computer software that made it much easier to produce fake photographs, we were, from 1988 onwards, treated to an orgy of sensationalized claptrap illustrated with made-up pictures of the Houses of Parliament sticking out of the floodwaters, vines growing up the wall of Buckingham Palace, and palm trees fringing Blackpool beach. (Oh, we've done that on our own cover, have we? Tut, tut.)

In this chapter I've collected together columns written at various times since that first short piece in the long-defunct *Today* newspaper almost 20 years ago, but because most of them were triggered by a particular news story, or politician's comment, I've dated them all and added a brief explanatory introduction. Careful readers may also be able to detect my own learning curve on the subject, and, I hope, a degree of consistency in my intolerance of sensationalist and unscientific opinions at both extremes of the debate.

Does carbon dioxide affect the climate?

From the Today *newspaper, April 1986*

Perusing one of the very earliest issues of *Weather* magazine, one of the journals produced by the Royal Meteorological Society, I was interested to see, amidst titles such as *Meteorology in Grammar Schools* and *The Origin of the Fahrenheit Temperature Scale,* as well as job adverts for weather forecasters in Nigeria and the Gold Coast, an article headed *Can Carbon Dioxide Influence Climate?*

It may surprise some people that our climate has been changing, admittedly relatively gently, since the end of the last ice age. Even nineteenth-century meteorologists thought that our climate had become more-or-less stable, and that is why they considered that thirty years of weather records was sufficient to describe the climate of every location on the Earth's surface. It may surprise even more people that it is well within the capability of human beings to cause such uncontrolled changes in the world's climate and that the results could quite possibly be catastrophic.

Although the article in *Weather* was written by G. S. Callendar in 1949, the theory goes back much further. Callendar himself had presented a scientific paper to the Royal Meteorological Society on the same subject in 1937, but in the 1949 piece he provided some historical background. It appears that an interpretation of climatic change in terms of the varying quantity of carbon dioxide in the Earth's atmosphere was first proposed in the 1880s by the Swedish physicist Sevante Arrhenius who conducted a series of experiments on the differing absorption rates of heat energy by a variety of gases.

After an initial flurry of interest among Arrhenius' contemporaries, the theory was largely neglected between the 1890s and 1930s. During this period the importance of water vapour in the lower atmosphere as a barrier to the radiation of heat from the Earth's surface to outer space was recognized and quantified. But during the 1930s more precise measurements showed that carbon dioxide (and ozone, too) played a significant part in the radiation balance of the planet.

Callendar was essentially a practical scientist – an applied physicist to be more precise. He was not satisfied simply with theorising. Developing theories in the laboratory was all very well, but he regarded the testing of those theories to be an equally important part of the scientist's job.

It is interesting that Callendar predicted that the rising proportion of carbon dioxide in the atmosphere would result in a rise in global temperature, but he did not regard this as a bad thing. His analysis was that 'there are reasons to expect some slight amelioration in climate to follow from the increase [in carbon dioxide]'. Callendar proceeded to examine temperature records from several long-standing observing sites in Europe, the Americas, Asia, the Pacific, and the Arctic. All of these exhibited an upward trend in annual mean temperature stretching back over seventy years. It is clear also that, even four decades ago, researchers were well aware that these trends might be affected by the growth of urban areas which can have a warming effect on the lowest part of the atmosphere under certain conditions, also by changes in instrumentation and observational procedure, and therefore the necessary allowances were made even then.

Callendar's conclusion was a model of measured equivocation: '. . . the climates of the world are behaving in a manner which suggests that slightly more solar heat is being retained in the atmosphere. This could be due to its increasing opacity to terrestrial heat as a result of the addition of carbon dioxide . . . but the coincidence itself is no proof.'

The green nightmare: Alpine skiing season of 1987–88

The Sunday Telegraph, *January 1988. Asked whether the snow-free slopes of Christmas 1987 were the first indication of what global warming might do to the Alpine winter holiday season, this is what I wrote.*

Two weeks of warm sunshine in December wrecked Christmas skiing holidays for thousands of visitors to Alpine resorts. And as December 1987 gave way to January 1988, newspapers and televisions screens were plastered with pictures of green slopes and browned off tourists. So it is with more than a little anxiety that prospective holiday-makers are beginning to scour the weather reports from Switzerland and Austria and France to spot any sign of the snows of winter '88–'89. Not just holiday makers, of course. There are plenty of executives from package holiday companies who are on their knees praying for a snowy January and February.

The main message we get from the disastrous start to the season is, never trust December. 1987 was not the first year that Christmas snow failed to materialize in the required amounts. December weather in Europe was just as unfavourable in 1980, 1975 and 1972, but it has only really been in the last few years that the Brits have flocked to the Alps in great numbers so early in the season. Old hands will tell you, quite correctly, that there is almost always much more snow at Easter than there is at Christmas, and besides, the days are longer, the sun is higher in the sky, and the wind is not as cold.

One question I am asked with very great frequency these days is how the so-called greenhouse effect is going to affect our weather. And it is only a short step from there to ask if it is going to affect the weather in the winter sports resorts in Europe. The greenhouse effect, remember, is the gradual rise in temperature caused by increasing amounts of carbon dioxide in the atmosphere, and the first signs of that warming process are now being measured.

The big difficulty is that it is not as simple as it sounds. For a start, scientists disagree about how much warmer the planet will become during the next fifty years; some say that the temperature increase may be just 1 degC (about 2 degF) while others estimate a possible warming of up to 4.5 degC (8 degF). That of course would be catastrophic, leading to the melting of parts of the polar ice caps, which in turn would produce a significant rise in sea level which in turn would threaten countries with large areas of land close to sea level such as Bangladesh, the Maldives, and even the UK. Day to day weather patterns would also change dramatically.

But the experts cannot yet say what the new weather patterns would be. They are all agreed that the temperature increase would not be the same all over the world, and most suggest that the polar regions would warm up more than the tropics. It is even possible that some relatively small areas would actually grow slightly colder.

As far as the winter resorts of the Alps and Pyrenees, the Dolomites and the Balkans, it is not necessarily bad news. Lack of snow in Europe is more usually a result of prolonged dry weather than of it being too warm. Occasionally, like this winter, you get both. But the warmer the air becomes the more moisture it can support, so a possible scenario for the European mountains once the greenhouse effect takes hold is this. Temperatures rise a few degrees, the moisture content of the air increases, more clouds form producing heavier rains at low levels and heavier snows at high levels. The increased cloud cover also reduces the amount of sunshine, so although temporary thaws may occur more frequently, they would not remove much of the snow. End result, a slightly shorter snow season but with less variation from year to year, and the upper slopes of the Swiss, French and Austrian Alps may finish up with more snow than they get now.

It must be emphasized that this is all speculation – nobody

knows for sure. But if the greenhouse effect lives up to the expectations of the most pessimistic of the climatologists we will all have a lot of other things to worry about than a spoilt skiing holiday.

A weatherman's view from the greenhouse

The Sunday Telegraph, *25 May 1990. The Editor of the day, Trevor Grove, asked me for an insider's view of the so-called greenhouse effect.*

Compare mad cow disease and the greenhouse effect. Two completely different subjects, I know, but look at it this way. On the one hand you have conservative, cautious government scientists saying that any danger is very remote, the minister acts accordingly, and half the country are baying for his blood. On the other hand you have conservative, cautious government scientists saying unequivocally that global catastrophe is nigh, apparently with a Prime Ministerial seal of approval, yet promised government action is minimal and the experts are widely accused of being merchants of doom.

Why the difference? You would probably need a United Nations panel of experts to determine that as well, but some of the reasons are quite obvious. There are, of course, enormous vested interests in the use of fossil fuels, and some of these have considerable influence on the governments in many countries, and there is also the logic that any individual government that takes unilateral action to cut carbon dioxide emissions puts itself at a commercial disadvantage (in the short term at least). Global warming is a global problem that requires global responses.

But this still does not answer the question why the response of the general public has been so much cooler towards what is immeasurably the greater threat. I can only suggest the following theory. For any individual, mad cow disease presents a small definable problem requiring a simple decision which threatens neither way of life nor standard of living. The greenhouse effect is a completely opposite kind of problem, and the natural human response is to emulate the ostrich and bury one's head in the sand and hope the problem goes away.

Some people do this by cherry-picking evidence and theories to support the contention that the atmosphere–ocean system can absorb any warming resulting from increased concentrations of carbon dioxide. It has been suggested that a global rise in temperature means that the atmosphere can hold more moisture, producing more clouds, thus reflecting the sun's energy back into space, which causes the temperature to fall again. That of course would explain why the extremely hot Sahara desert is so cloudy!

Even if our planet were capable of absorbing the effects of increasing the concentration of greenhouse gases, that is even more frightening than straightforward global warming. Such a scenario may offer a period of fifty, or even 150, years with little or no increase in global temperature; peoples and governments become complacent, and carbon dioxide emissions continue to rise. But the Earth's ability to absorb growing quantities of heat energy must be finite, eventually the stress on the system becomes intolerable, the atmospheric checks and balances fail, the pressure-cooker explodes, and a runaway greenhouse effect ensues. That would be immediately catastrophic.

Others say, 'Why are we worried about warmer weather? Let's take advantage of it. Let us look forward to the day when vineyards adorn the countryside as far north as Paisley and Perth, and Europeans visit our coastal resorts for their summer holidays. Canada and Siberia can become the new bread-baskets of the world.' The trouble with this is that *any* change of climate requires adaptation. People have to adapt, economies have to adapt, and the natural flora and fauna of a region have to adapt. If the climate changes at an unnaturally fast rate, plants and animals may fail to keep up. It is much easier for one particular variety of plant to die out in a given district than it is for that plant to establish itself in another district. Can you seriously imagine the great creaking Soviet agricultural bureaucracy coping with climate change? And what happens to the grain-growing capital of the world, the American Mid-West, as temperatures rise and rains fail? As for the UK, what is the point of having a Mediterranean climate if there are no beaches left, thanks to the rise in sea-level?

In my capacity as a professional meteorologist I am asked increasingly frequently what I think about the greenhouse effect. Like Government scientists I, too, am fairly cautious, and in the past I have usually ended up playing devil's advocate: disowning the more lurid scare stories, but at the same time accusing the head-buriers of ignoring the evidence.

Global warming *is* happening. Experts do not know how it will proceed; more important, nor will we ever know exactly how it is going to proceed until it has actually happened. Thus there is no point in governments waiting to decide their policy until the scientists come up with fireproof answers. The government of the USA has yet to admit to the existence of the greenhouse effect. This is no less than a complete abdication of its international responsibilities in the face of pressure from big business. What they do not seem to realize is that their country stands to lose more, financially, than any other nation in the world during the next 100 to 200 years.

This all makes me very pessimistic, cynical even, about our ability to cope with climate change over the coming decades. Governments have a habit of concentrating their efforts on four or five year stretches, for obvious reasons. Spending tax-payers' money on something that may not make its presence felt for fifty years is hardly an electorally popular pastime.

Thus I do not believe the majority of administrations around the world will do other than pay lip-service to the need to take action. Talking-shops will abound.

I suspect it will take major natural catastrophes in several individual countries to create the political will to follow the experts' advice. After all, the Clean Air Act reached the statute books in the UK in 1956 as a response to the Great London Smog of 1952 in which some 7000 people died. This followed six hundred years of ineffectual legislation on air pollution in London. Huge investment in sea-defences, including the Thames Barrier, followed the devastating North Sea floods of January–February 1953 when over three hundred died in the UK alone. Before 1953 our sea-defences had endured decades of benevolent neglect. I have a ghastly feeling that it will take a similar disaster before a British government – of whichever colour – perceives it electorally advantageous to tackle seriously the problems posed by global warming.

Global climate change and individual weather events

The Sunday Telegraph, *16 August 1998. Vice-President Al Gore made the mistake of linking an individual weather event in a small part of the world with global climate change. It is like comparing an apple with the fruit harvest.*

Mr Gore, the American Vice-President, muddied the waters when he linked the report from government scientists that July was the warmest month, averaged globally, this century with the continuing heatwave in southern and western parts of the USA. As climate experts never cease to point out, the existence of abnormally hot weather somewhere or other on the planet is not evidence of an underlying warming trend. Nor does a protracted spell of frost and snow indicate that any such trend has ended or perhaps never existed at all.

The report on July's global warmth reached the British

newspapers on Monday morning, which coincidentally turned out to be hottest day of the year (so far) in many parts of the UK. The temperature reached 32.2°C (90°F exactly) at Broadness, near Gravesend, in Kent, and 31–32°C (88–90°F) at several other places in London and the Home Counties. Thus it was hardly a surprise that a few of our own journalists tried to associate the hot spell with the world-wide warming trend; they perhaps had momentarily forgotten that the report referred to July whereas our little heat-wave only began on August 4.

One of the disadvantages of studying a subject in depth is that you become more and more conscious of your own limitations. After spending seven years at university, over twenty years reading, writing, and broadcasting about meteorology and climatology, and many opportunities to talk to the world's experts in these fields, I have become painfully aware of how little I actually know.

This is, of course, a problem unknown to most politicians. They read their sheet of A4 – a briefing note prepared by someone who understands only a little more than they do – and then they mouth off about the subject as if they were the nation's leading authority on it. It is embarrassing. The same goes for much coverage in the media. One editorial on climate change I read recently was so full of factual errors I wondered whether the author had ever read his own publication.

So what do I think about climate change? I know little enough about the subject to defer to the world's experts, and I know enough about it to realize that the biggest sceptics and the biggest scaremongers are motivated by a political agenda and their attempts to justify their positions are fallacious and are based on a misunderstanding or a misrepresentation of the science.

Common sense needed to understand global warming

The Sunday Telegraph, *18 October 1998. More of the same, two months later.*

Here we are again. Government reports on the latest research into global warming land on the minister's desk on the same day as the official statistics showing that September was, averaged over the entire planet, the warmest since useful global statistics began some 120 years ago. And the end result is a government spokesman parroting hypothetical scenarios from his briefing document as if they were facts.

Let us deal with those global statistics for September. Not only was it the warmest September to date, but every month so far this year has broken its previously established record. The chance of such a sequence occurring randomly is infinitesimally small, and the obvious conclusion to be drawn is that the global climate is becoming progressively warmer. An alternative explanation is that the statistics are erroneous.

Those who are blindly sceptical about global warming seize upon this latter idea, suggesting a range of reasons, including urbanization of recording sites (cities are warmer than the countryside), changes in instrumentation, changes in the way sea-temperatures are measured, and a lack of corroborative evidence from satellite data.

I do not subscribe to this view. It is arrogant of those who peddle these criticisms, and patronizing in the extreme to climatologists, to suggest that the scientists involved are not aware of these possible sources of error. The truth is that they have all been studied in detail and quantified so that any errors are eliminated. For instance, the contribution to the perceived warming trend from the growth of towns around weather-recording sites has been known about for over a century, and published statistics have taken it into account for many decades.

What does worry me, however, is the way government min-

isters and the news media present the findings of the experts to the general public. The climatologists who write reports about global warming use words very deliberately. They are laced with phrases like 'one possible scenario', and 'is perhaps a more likely outcome', and 'is indicated if present trends continue'. Mr Michael Meacher on the radio on Friday did not use these phrases; instead he used 'will', 'will', and 'will'.

The government is right to make the latest research of its experts available to us. But it is wrong to present considered opinion as fact. It is right to tell us what we as a nation can do to reduce emissions of pollutants, but our contribution will be a fat lot of use if others – especially the Americans – do not follow suit.

A warmer world does not necessarily mean a warmer Britain

The Sunday Telegraph, *25 October 1998*

Recent reports by climatologists, both in government service and in the universities, have concluded that a continuing warming trend in the global climate is likely to be matched by a warming trend in the climate of the UK. But there remains considerable uncertainty about how the changing climate at a global level will be reflected at a regional level. Although conventional wisdom in the late 1990s is that our own climate is more likely than not to become warmer, it is by no means a foregone conclusion.

A look at a couple of meteorologically dramatic years will serve to illustrate this. The summer of 1976 was the hottest and driest in Britain for over two centuries and contained a truly exceptional heatwave during which 32°C (90°F) was exceeded somewhere in the UK on fifteen consecutive days. The longest such spell before or since has been just five days. Although the later months of that year were rather cold, the mean temperature for the whole of 1976 in the UK was higher than any other year between 1960 and 1988. Averaged globally, however, the only *colder* years than 1976 in the last half-century were in 1964 and 1956. By contrast, 1963 was one of the coldest years of the twentieth century in Britain and it brought us our most severe winter since 1740, but globally it was the *warmest* year of the 1960s.

The suggestion that the UK might benefit from a quasi-Mediterranean climate by the middle of the next century has, understandably, gained a lot of coverage in the media. Superficially, such a change might be considered a good thing. But what is the use of beach-friendly summers if our beaches are washed away by the rising sea level? What's the use of vineyards in Yorkshire and paddy-fields in Somerset if we can no longer grow wheat and barley? What's the use of lower heating bills in winter when you have to pay through the nose for air-conditioning units in summer? New pests and bugs have already crossed the Channel during recent hot summers, and this process may accelerate. Photochemical pollution (especially higher ozone levels) would also accompany a rise in the frequency of hot summers.

Our social and economic structures, particularly the highly complex ones of the western world, are closely attuned to our environment. If the environment changes, those structures have to be adjusted, or they collapse. In less advanced parts of the world climatic change has historically led to mass migration, and a prolonged warming trend in the climate of these regions will increase the likelihood of creating millions of 'climatic refugees'.

Kate Adie and global warming

The Sunday Telegraph, *14 February 1999. Another journalist commits the sin of linking a spell of unusual weather with global warming.*

I could scarcely believe my ears this (Saturday) morning when I heard Kate Adie, introducing 'From Our Own Correspondent' on Radio 4, say that scientists were blaming the heavy snow in the Alps and last week's avalanche at Chamonix on global warming.

Oh, puhleeze!

Are these the same scientists who, after the virtually snow-free Alpine winters of 1988–89, 1989–90, 1991–92 and 1994–95, were warning us that global warming meant much less snow in the Alps in future decades? The European winter holiday industry, they said, would have to up sticks to Scandinavia.

I hesitate to re-visit this subject. It is rather like swimming against the tide. Every time we have an unusual weather event, some clever-dick will suggest it has something to do with global warming. Somehow, I doubt any reputable scientist was behind last week's comment linking the French avalanches with the changing climate. More likely it was some self-styled commentator who thinks that reading *New Scientist* every week qualifies him as an expert.

The common-sense view is simple. There is a warming trend in the planet's climate, part of which is probably due to the increasing emission of greenhouse gases into the atmosphere. But we do not yet know what proportion is due to this enhanced greenhouse effect, and what proportion is explained by the natural variation of the world's climate. Unusual or dramatic weather events have always happened, and we do not need to point the finger at global warming to explain them. The same is true of exceptional or extreme months and seasons.

It is a logical fallacy to argue from the particular to the general. In the same way, the fact that the Mayor of Chamonix says that there had not been such an avalanche in this part of the valley since 1908 does not, as some media pundits suggested, mean that this was the worst avalanche in the Alps (or even in Chamonix) for ninety years. Five minutes with my history books revealed that, for instance, two major avalanches in early 1970 in the French Alps – in Val d'Isère and St Gervais – killed 114. And in 1951, a much snowier winter than this one, a series of avalanches in Switzerland during February killed almost one hundred.

Please, Kate, the next time you have some unusual weather to report, see if you can do it without mentioning global warming.

Politicians ask butterflies if the climate is changing

The Sunday Telegraph, *27 June 1999*

Fruit blossom, butterflies and ski slopes have one thing in common. They are all on the list of things that government has decreed should be monitored in order to understand how our climate is changing.

There is nothing new in observing all God's creatures to help predict the weather. The Americans, as all regular film-goers will know, have Groundhog Day. This falls on 2 February and is supposed to hold the key to the weather for the rest of the winter. If the groundhog – *Marmota monax* – emerges into sunshine sufficiently strong to cast a shadow at noon, it will return underground for six weeks and the rest of the winter will be cold and frosty. Few Americans realize that this legend is a European import, 2 February (or Candlemas Day) having more old country weather lore associated with it than any other date in the calendar.

Plants and animals are, of course, affected by the weather, but they are affected by other things too. Information collected about flowering dates, hibernation times, the arrival and departure of snow and ice, and so on is called 'secondary data' or 'proxy data' by climate experts.

I am increasingly perplexed when I hear politicians and some commentators talk about these proxy data as though they give unequivocal evidence of changes in the climate, whereas proper climatological records of temperature, rainfall, sunshine and wind are, to them, untrustworthy. I have no problem with the idea that these botanical and zoological indicators are providing us with additional information about the way the UK's environment *responds* to our changing climate. They most certainly do. But that is very different from *measuring* changes in the climate itself. All these secondary indicators are affected by other factors such as human intervention, genetic modification, pesticides, fertilizers, pollution, and complex interactions with other elements of a changing ecology. Some of them respond to climate change with a delay of years or even decades.

Those who study the history of climate change – the palaeoclimatologists – use this sort of proxy data to reconstruct past climates before thermometers and rain-gauges were invented. Other climate markers such as the pollen contained in lake and ocean deposits and oxygen isotopes trapped in the ice-caps of Greenland and Antarctica help these experts understand the climate of the ice age and before. They know all too well how limited, and sometimes contradictory, the information can be.

Antarctic melting and bad journalism

The Daily Telegraph, *11 December 1999. A series of presentations by an environment correspondent on BBC news programmes demonstrated how easy it is for a non-expert to misunderstand completely a scientific report. This measured report was published by the British Antarctic Survey, but the television news stories promised a complete melting of the Antarctic ice before the end of the twenty-second century and the inundation of most of the world's great cities, including London.*

A sea-level rise of seventy metres (230 feet) in ten generations: that, according to some of the tabloids and even some radio and television news bulletins, is what will happen because the warming trend in the global climate will bring about the complete melting of our planet's ice-caps.

This is essentially a re-run of the outrageous stories which appeared when the media first discovered global warming ten years or so ago. Maps showing the remnants of the British Isles were typically accompanied by pictures of the dome of St Paul's Cathedral poking through the waves. It is also incompetent and exaggerated reporting allied to an editorial predilection towards sensationalism. Trying to explain climate change in an objective and scientific way is difficult enough without having to spend time countering this sort of disgraceful journalism. Scientific researchers who speak to the media are sick to the back teeth of being misquoted and having their comments gratuitously edited to lend credence to the sensationalized treatment.

Last week's story revolved around work currently being carried out by the international Cape Roberts research team in Antarctica. So far, they have confirmed the already known fact that there was no ice in Antarctica before roughly 35 million years ago, and that ice began to form there when the mean global temperature was 5 to 6 degC (around 10 degF) higher than it is now. We also know that several climate prediction models indicate that the planet may experience this sort of temperature rise during the next 200 to 300 years if there is no reduction in the emission of so-called 'greenhouse' gases.

The fallacy is to infer that a 5 to 6 degC temperature rise *now* will result in a return to the ice-free environment of 35 million years ago. The first thing to point out is that, notwithstanding the much higher global temperature then, it was actually cold enough for the ice-cap to start developing around the South Pole. Secondly, even if the planetary climate patterns were to change sufficiently to lift polar surface temperatures above freezing, it would take many, many thousands of years for all the ice to melt.

But, crucially, it also ignores the changed geography of the world resulting from continental drift. The land link between Antarctica and South America had, 35 million years ago, only recently been broken, while Australia also lay much closer to the polar region than it does now. Thus, compared with today, the Southern Ocean was much narrower, the circumpolar ocean current was much weaker, and the atmospheric circulation was also much less vigorous. Furthermore, the Arctic Ocean was broader and open to warm oceanic currents from the Pacific, which is not the case today. All these factors favoured a warmer world than now. Glaciation of Antarctica began when the Southern Ocean broadened, isolating Antarctica meteorologically from the rest of the southern hemisphere. Today, the Southern Ocean is broader still, Antarctica remains isolated meteorologically from the rest of the planet, and the global climate will continue to favour a glaciated southern continent for many centuries. For Antarctica to lose all its ice now it would require a massive global temperature rise: by the time it was warm enough for the Antarctic ice to melt the rest of the planet would probably have become so hot it would be quite uninhabitable.

How volcanoes affect the climate

The Daily Telegraph*, 25 March 2000. Several volcanic eruptions, some more violent than others, had occurred during the early weeks of 2000. The Features Editor asked me if volcanoes and earthquakes were caused by the weather. I bit my lip, said 'no', but then went on to explain how volcanic eruptions could change the climate. 'OK,' she said, 'do that, then.'*

Our planet seems to be suffering a bout of indigestion at the moment. During the last few weeks several well-known volcanoes at widely scattered locations across the Earth's surface have been grumbling, including Etna in Sicily, Hekla in Iceland, Soufrière on Montserrat, and Mayon in the Philippines.

It has been known for many years that major volcanic eruptions can have a significant impact on the global climate. Indeed, the link between volcanic activity and the weather in a given region – for instance in western Europe following the eruption of Laki in Iceland in 1783 – was made by scientists centuries ago although they were unaware of the global climatic response.

The mechanism by which the biggest eruptions affect the world's climate is now fairly well understood. The ejection of debris and gases needs to reach the stratosphere to have the greatest impact, for this is a region of the atmosphere, more than ten kilometres (six miles) up, which is above the weather, so the dust and other material is not 'rained out'. Just as important, stratospheric winds blow in such a way that ejected matter from a volcano close to the equator is gradually spread across the planet to form a veil which lasts on average for about three years, and very occasionally for as long as seven years or even more. The dust veil reduces the strength of the sun, but does not limit the amount of heat energy which escapes from the Earth to outer space at night and during the winter.

The solid particles in the dust veil gradually drop out of the stratosphere under the influence of gravity whereas gaseous matter lasts much longer. It is now known that eruptions with a high sulphur content have the greatest impact on climate. The sulphur reacts with water vapour in the atmosphere to form billions of tiny droplets of sulphuric acid which are much more effective than ordinary dust at diminishing the energy coming from the sun.

All this tells us that the greatest cooling effect comes from a volcano in tropical or equatorial latitudes (which lets out Hekla and Etna) which erupts explosively to send its plume of debris high into the stratosphere (which lets out Soufrière), and which has a high sulphur content. Mayon has not yet ejected sufficient sulphurous material into the stratosphere to have a significant effect, but it may yet do.

The last climate-changing eruption was that of Pinatubo, also in the Philippines, in 1991, and before that, El Chichón in Mexico in 1983. It is calculated that the global temperature dipped by between 0.5 and 1.0 degC (1 to 2 degF) during subsequent years as a result. The Central England temperature in 1992 was 0.4 degC below, and in 1993 0.7 degC below, the average for the rest of 1990s. Many readers will remember the hazy brown ring that encircled the sun during those two summers – a direct consequence of the veil of sulphuric acid droplets in the lower stratosphere.

The four pillars of climate change

The Sunday Telegraph, *8 October 2000*

Scientists at the heart of research into global climate change have great difficulty in explaining their results to the public at large because the whole subject has been politicized. This is hardly surprising: the findings of the world's leading atmospheric science experts and the predictions they make concerning future climate change all demand political decisions.

Few of these scientists have a political agenda, and even fewer are politically experienced. They dislike giving interviews and they are deeply frustrated by the way their statements are partially quoted and their scientific papers are cherry-picked by those who are making the political arguments.

Before you pick up your pen to complain about my own partiality, allow me to let you into a little secret. When I have addressed this subject before, my correspondence has divided into roughly four equal parts – those who want me to resign for falling for the propaganda of the environmental pressure groups, those want me to resign for falling for the propaganda of the so-called 'denialists', those who sing my praises for seeing the sense of the environmentalists' arguments, and those who sing my praises for seeing the sense of the denialists' arguments. If you wish to add to one of those four piles, that of course is up to you, but I should really add one more thing. The more I learn about the way our planet's climate works (quite a lot, I guess, after more than thirty years studying, reading and talking, from University days to membership of the Council of the Royal Meteorological Society), the more I realize how little I really do know, and it continues to surprise me that some people who clearly know much less still seek to lecture me.

I hear axes grinding loudly at both extremes of the argument, but last week saw the publication by the World Wide Fund for Nature of a report entitled 'Climate Change and Extreme Weather Events' which purported to show that the evidence that recent weather extremes were the result of global warming was 'overwhelming'. No reputable scientist would put his name to such a statement. There is certainly an argument that warming the Earth's atmosphere is likely to change the frequency and intensity of extreme events, but, crucially, recent weather disasters fall well within the envelope of past extremes. That these disasters are having a disproportionate effect on human populations is undoubtedly true, but that is a different argument.

The hottest place in the world

The Daily Telegraph, *12 May 2001. A noted American climate-change sceptic had been quoted in the press a few days earlier saying that, if global warming were true, one of the first records to be broken would have been the world's highest individual temperature. As this had stood for almost 80 years, he said, it was clear that the alleged warming trend in the planet's climate was nothing but a scare story.*

If you believe all the reference books, the hottest place on our planet is a small Libyan settlement called al-'Aziziyah which is located just over thirty kilometres (almost twenty miles) inland of the capital, Tripoli, on the road which runs south into the Sahara desert.

In 1913 a group of geologists working for the US National Geographic Society in al-'Aziziyah established a weather-recording station there, and on September 13, 1922, a temperature of 57.8°C (136°F) was recorded there, and this is the figure that you will find in the record books. However, some controversy surrounds it, and it is not accepted by the Libyan authorities. Now, one might conjecture that such a rejection is merely the predictable response of a régime which despises all things American, even weather observations. In fact it dates back to the time the record was set; Libya was then occupied by the Italians and the Director of the Libyan Meteorological Service at the time, Dr A. Fantoli, found that the reading was unacceptably high. Even at a distance of eighty years we can find evidence to support the record, and also reasons to reject it.

The geographical setting of al-'Aziziyah favours abnormally high readings under the right meteorological conditions. It lies to the lee of a steep escarpment some 600 metres (2000 feet) high which marks the northern flank of the mountain-range known as the Jabal Nafusah. When the wind blows strongly from the south or southwest, from the interior of the Sahara desert, the air is forced to climb over the Nafusah hills whence it descends to the Gefarah plain. During this descent the air is warmed by compression, and if the air had originally been moist enough to deposit some rain over the high ground (very rare but by no means impossible in this part of the Sahara) that warming process would be particularly efficacious. Such a process is the same as is produced by the 'föhn' wind of the Alpine region or indeed the 'föhn effect', which our own weather forecasters occasionally talk about when a warm moist wind blows across the Welsh or Scottish hills.

On the day in question there was certainly a southerly wind

blowing, but there is no evidence even of extensive cloud, let alone any rain, over the Jabal Nafusah. One would also expect, if a föhn effect were operating, that abnormally high temperatures would have also been recorded at neighbouring stations, but at Tripoli the day's maximum was 45°C (113°F) – a huge difference. The fact that Tripoli lies on the coast adjacent to the cooler waters of the Mediterranean Sea is irrelevant in this case because the wind blew strongly from the Saharan interior throughout the day.

When we probe deeper among the records we discover some fascinating discrepancies. Before the Second World War there were three weather stations in the area: Tripoli, where readings were made from 1879 to 1939, al-'Aziziyah (1913–1940), and, roughly half way between the two, Castel Benito at Idris (1924–1941). During this period al-'Aziziyah's hottest days averaged 6 degC (11 degF) higher than those at Idris and 9 degC (16 degF) higher than Tripoli's. When the stations re-opened after the war, al-'Aziziyah averaged 1 degC (2 degF) lower than Idris and 2 degC (4 degF) higher than Tripoli in extreme conditions.

It is impossible to be dogmatic about the reasons for this enormous change, but it is consistent with a faulty thermometer, a damaged thermometer-screen, an over-sheltered site, or a combination of all three. Whatever the reason, the records are so inhomogeneous that they simply cannot be trusted. On this one I am with the Libyan authorities!

The snows of Kilimanjaro

The Daily Telegraph, *27 October 2001. This column followed a story on the news wires, picked up by several newspapers, which claimed that the retreating snows on the upper slopes of Kilimanjaro were clear evidence of global warming.*

Snow falling at the equator? As an eight-year old swot, convinced that I was a better geographer than Mr Evans, my class teacher at William Austin primary school, I knew that snow at the equator was quite impossible. I may have heard of Kilimanjaro and Mount Kenya but I certainly did not realize, as Mr Evans did, that they were snow-capped.

In my childish ignorance I was in good company. My reaction was exactly the same as that of early Victorian geographers who described rumours of snow-capped mountains in the middle of the Dark Continent as 'preposterous'. Mount Kenya is located just 16 kilometres (ten miles) south of the equator and is sometimes visible from Nairobi just 110 kilometres (seventy miles) away, and its summit is 5200 metres (17,057 feet) above sea level. Kilimanjaro is 320 kilometres (200 miles) south of the equator and rises 5895 metres (19,340 feet) above sea level. Although daytime maxima in the coastal lowlands of Tanzania and Kenya are routinely between 27 and 32°C (80 to 90°F), we now know – unlike nineteenth century geographers – that the temperature falls with altitude at a rate of about 1 degC (1.6 degF) per hundred metres (330 feet); that places the uppermost slopes of both of these spectacular mountains above the freezing level for much of the year.

Kilimanjaro was first climbed by Europeans in 1889 by Hans Meyer of Germany and Ludwig Purtscheller of Austria, led by a local guide. Twenty years later it was fully surveyed by the colonial authorities – the Tanzanian mainland at that time was known as German East Africa.

By comparing the results of that survey with present-day aerial photographs of the peak, American glaciologist Lonnie Thompson has determined that 82 per cent of the snow and ice has disappeared in the last eighty years. Other surveys in the 1940s and 1950s indicate that this has been a progressive change and not a sudden recent one. The inference is that there is a long-term climatic trend at work here, and, although the slight acceleration in the rate of melting in the last twenty years may be associated with recent anthropogenic climate change, it would be wrong to place all the blame for the disappearance of Kilimanjaro's snows on so-called 'global warming'. We can get an even better perspective from the writings of the Portuguese traveller, Manoel de Almeida, who noted a snow-cover above 3600 metres (12,000 feet) in the Ethiopian Highlands, which learned elders in Addis Ababa told him never disappeared. There are no permanent snowbeds there now. If the trend continues, neither Kilimanjaro nor Mount Kenya will have any glaciers left by 2025.

We should be wary of making a direct link between the changing pattern of snow-cover on the one hand, and temperature levels on the other. Nor should we necessarily assume that temperature variations at high altitude mirror variations at sea level. There are several other factors involved, including precipitation rates, cloud-cover and sunshine, humidity levels and wind. Put at its simplest, if snow does not fall, any snow on the ground will gradually disappear by ablation and sublimation no matter how low the temperature is.

Analysing past changes in the climate

The Daily Telegraph, *19 January 2002. One of the questions most frequently asked over the years in respect of climate change is, 'How do you know what the climate was like before barometers, thermometers, and so on, were invented?' This was an attempt at providing an answer.*

The question that immediately springs to mind when the boffins talk about severe winters in the Middle Ages, let alone wet years in the Bronze Age, is: how can they possibly know? After all, thermometers and rain-gauges were only invented in the seventeenth century, and systematic recording of the vagaries of the weather only began in the middle of the eighteenth century and then only at a few isolated locations in Europe.

Before the instrumental period a completely different approach is needed, and this involves bringing together a wide variety of what are called 'proxy' data. Climatologists have collaborated with historians and economists, geologists and geographers, botanists and zoologists, vulcanologists and chemists, to set out upon this massive task of reconstructing past climates.

Ships' logs include invaluable records of weather on regular shipping routes back to the middle of the seventeenth century with intermittent and localized data from much earlier than that. Similarly, the habit of keeping weather diaries was quite common amongst the middle classes – clergymen, doctors, politicians, for example – during the sixteenth and seventeenth centuries, not just in Europe but in the American colonies too, and one outstanding register kept by an English priest, William Merle, dates back to 1337–1344.

General journals, including those of famous diarists such as Samuel Pepys and John Evelyn, also contain a wealth of descriptive information, especially of extreme weather events or exceptional seasons, while economic historians are able to extract a lot of climatic information from farm and estate records, dates and yields of harvests and vintages, the price of foodstuffs and other essentials, dates of freezing and re-opening of rivers and harbours, and even records of military campaigns. Taken together, this amazing variety of information has allowed climatologists to complete a chronicle of important climatic events across much of Europe back to 1066, and sporadic information has been gleaned for earlier periods too, particularly during Roman times.

Tree rings reveal significant climatic information including the length of the growing season and rainfall during spring and summer. Overlapping records – recently cut trees, timbers in old buildings, tree trunks recovered from bogs and lakes where the water prevents decomposition – extend our knowledge back to 2000 BC in parts of Europe and to 6000 BC in the south-western USA. Similar techniques are used to build a chronology of mud deposits in lakes and river estuaries, revealing information about river flow which is in turn dependent on rainfall, and in Sweden sediment records beginning almost 10,000 years ago have already been analysed. The ice-sheets of Antarctica and Greenland also lend themselves to this sort of investigation, with snowfall data extending back 250,000 years.

Zoologists and marine biologists help climatologists to study insect populations, pollen distribution, and the remains of tiny sea-creatures, all of which are climate-dependent in their geographic distribution, and all of which have been preserved in sedimentary deposits. These are, generally-speaking, a fairly crude method of ascertaining broad changes in the climate, but they do allow us to fix major climatic shifts in the last million years to an accuracy of better than 100 years in some cases.

Climate change 1940s-style

The Daily Telegraph, *16 February 2002*

Fifty years ago climate historians were, in that restrained fashion typical of academics and government scientists of the day, rather excited by the curious behaviour of the British climate during the 1940s and early 1950s. Nor was the unusual weather confined to the British Isles: changes were being noted in many different parts of the world.

Until 1940, the twentieth century had shown a gradual warming trend in all seasons in the UK. Indeed, long-term statistics showed that this trend had begun in the 1880s. There was no truly severe winter in Britain between 1896 and 1939, although those of 1917 and 1929 provided lengthy spells of frost and snow, whereas there were several exceptionally mild ones. Long hot summers appeared regularly between 1911 and 1935 and there was not a single summer that was significantly colder than average between 1925 and 1945 inclusive.

Then, all of a sudden, we had three successive very cold and snowy winters in 1939–40, 1940–41, and 1941–42, quickly followed by another snowy winter in 1945, and a memorably severe one in 1947. December 1950 was also very cold. In the space of twelve years we had endured six winters which were as cold, or colder, than anything that had happened in the previous forty years. The springs of 1943 and 1945 were among the warmest on record, the autumn of 1949 was warmer than any before, while summer temperatures fluctuated wildly, the exceptional heat of 1947 and 1949 contrasting with the persistent coolness of 1946 and 1948.

The more suggestible in the climatological community trumpeted a significant change in the climate. The less impulsive divined what they called a 'climatic perturbation'. The reality was somewhere between the two: the general worldwide warming of the previous sixty years was clearly arrested in the 1940s and was followed by a general cooling which lasted until the early 1970s. This thirty-year reversal was also quite apparent in the British record. That cooling trend came to a clear-cut end around 1973 – ironically the year when one or two publicity-minded popular scientists hit the headlines with dire warnings of a rapid decline towards the next ice age. This imminent ice-age theory was a hypothesis that was not espoused by climatologists in the UK, but even today they have it thrown in their faces as an example of scientific sensationalism when the threat of future climate change is discussed. 'You cried "wolf" once before,' they are told.

Summers are getting drier

The Daily Telegraph, *7 June 2003. Following the previous summer's downpours, and at the beginning of summer 2003, I wanted to put on record some statistics which showed a dramatic decline in summer rainfall over much of the UK since the late 1960s. I could hardly have imagined the extent to which August 2003 would support my argument.*

The suggestion that the English summer is drier than it used to be may raise a few hollow laughs around the country. The statistics, however, reveal a substantial drop in rainfall over practically the whole of England, and also large parts of Wales and Scotland, during recent decades compared with the first two-thirds of the twentieth century. There has not really been a clear downward trend; rather, a step-change happened around 1970.

The words 'climate change' are these days often used as a synonym for the present warming trend, but the fact is that Britain's (and the world's) climate is always changing. As well as the remarkable warmth of the 1990s and the drying out of the English summer, last century also brought a marked increase in March rainfall, a substantial improvement in sunshine in November and December, a period of warm Aprils in the 1940s and 1950s, and period of cold winters between 1939 and 1987.

The drop in summer rainfall has been confined to July and August. Averaged over England and Wales June has actually become wetter during the last one hundred years: the mean rainfall for the month for 1911–40 was 57 millimetres (2.26 inches), for 1941–70 it was 61 millimetres (2.43 inches), and for 1971–2000 it was 68 millimetres (2.68 inches). However, the early part of the last century produced a run of unusually dry Junes, and the increase in rainfall since about 1950 has not yet returned this month to the level of wetness which prevailed for most of the nineteenth century.

The drying out has been most marked in July. Mean rainfall for England and Wales for 1911–40 was 83 millimetres (3.26 inches), for 1941–70 it was 73 millimetres (2.90 inches), and for 1971–2000 it was just 57 millimetres (2.24 inches), representing a drop of 31 per cent over sixty years. For the first time in 300 years of rainfall recording, July is now the driest month of the year, making a dramatic contrast with the period 1781–1830 when it was, alongside October and November, the equal wettest month of the year. The figures for August were 81 millimetres (3.19 inches) for 1911–40, 91 millimetres (3.60 inches) for 1941–70, and 73 millimetres (2.88 inches) for 1971–2000, revealing a 20 per cent drop between the last two periods.

The decline in mean monthly rainfall during July and August is supported by a similar drop in the number of days with rain, a drop in the frequency of very wet days, a decline in the frequency of thunderstorms, and an increase in the number of long rainless periods. All of these changes have been most marked in eastern and southern counties of England, but they are discernible in the Midlands, the West Country and northwest England as well. This has repercussions for agriculture, especially in parts of eastern England. Over a large part of Essex and Cambridgeshire irrigation is now required in at least nine out of ten summers.

Index